Of Thorn & Briar

Of Thorn & Briar

A Year with the West Country Hedgelayer

PAUL LAMB

**SIMON &
SCHUSTER**

London · New York · Amsterdam/Antwerp · Sydney/Melbourne · Toronto · New Delhi

First published in Great Britain by Simon & Schuster UK Ltd, 2025

Copyright © Paul Lamb, 2025

Illustrations © Robin Mackenzie, 2025

The right of Paul Lamb to be identified as the author of this work has been asserted in accordance with the Copyright, Designs and Patents Act, 1988.

3 5 7 9 10 8 6 4 2

Simon & Schuster UK Ltd
1st Floor
222 Gray's Inn Road
London WC1X 8HB

For more than 100 years, Simon & Schuster has championed authors and the stories they create. By respecting the copyright of an author's intellectual property, you enable Simon & Schuster and the author to continue publishing exceptional books for years to come. We thank you for supporting the author's copyright by purchasing an authorized edition of this book.

No amount of this book may be reproduced or stored in any format, nor may it be uploaded to any website, database, language-learning model, or other repository, retrieval, or artificial intelligence system without express permission. All rights reserved. Inquiries may be directed to Simon & Schuster, 222 Gray's Inn Road, London WC1X 8HB or RightsMailbox@simonandschuster.co.uk

www.simonandschuster.co.uk
www.simonandschuster.com.au
www.simonandschuster.co.in

Simon & Schuster Australia, Sydney
Simon & Schuster India, New Delhi

The author and publishers have made all reasonable efforts to contact copyright-holders for permission, and apologise for any omissions or errors in the form of credits given. Corrections may be made to future printings.

Simon & Schuster strongly believes in freedom of expression and stands against censorship in all its forms. For more information, visit BooksBelong.com.

A CIP catalogue record for this book is available from the British Library

Hardback ISBN: 978-1-3985-3503-9
eBook ISBN: 978-1-3985-3504-6

Typeset in Bembo Std by Palimpsest Book Production Limited, Falkirk, Stirlingshire

Printed and Bound in the UK using 100% Renewable Electricity at CPI Group (UK) Ltd

The authorised representative in the EEA is Simon & Schuster Netherlands BV, Herculesplein 96, 3584 AA Utrecht, Netherlands. info@simonandschuster.nl

This book is dedicated to my father, Anthony Lamb.

CONTENTS

1:	August	1
2:	September	17
3:	October	37
4:	November	57
5:	December	77
6:	January	99
7:	February	127
8:	March	155
9:	April	185
10:	May	213
11:	June	243
12:	July	271

Acknowledgements   295

1

August

For more than five years now I have been living in the wagon, travelling the lanes of the West Country between little-known farms and smallholdings where I ply my trade as a hedgelayer. Here, on lonely acres nestled well off the beaten track, I have felt most at home, surrounded by our countryside and its inhabitants. I am content with my nomadic lifestyle, moving from farm to farm, job to job, settling for some months only to move on again as work and the season dictate. In recent years my contracts have taken me through the rolling countryside of Devon and into the flat lands of Somerset to the north. I have laid hedges across the chalk downland of Dorset and Wiltshire away to the east, and down in Cornwall in the far south-west.

I travel between contracts in an ageing horsebox that is also my home throughout the year. This wagon contains everything I need to live comfortably and carry out my work. A large wooden trunk is housed beneath the body of the lorry and has more than enough room to store the paraphernalia associated with hedging work: chainsaws, fuel and oil, billhooks and axes, all essential articles in the hedger's armoury. The cab of the old lorry can seat four people comfortably, though I rarely accommodate passengers. Behind the beige vinyl seats is a broad parcel shelf made of plywood with ample space to store all of my outdoor equipment. A hand-forged kettle iron and chain span the width of the shelf whilst a trivet made of three horseshoes and an old metal ammunition box containing files, spanners and other small tools, are tucked neatly behind the driver's seat. Three jerrycans are stowed at the other end of the shelf and are kept filled with water. There are no storage tanks or water pumps on the wagon and I have to collect water from the farm tap, or nearest supply, and store it until needed. I keep two enamel jugs inside the wagon, each capable of holding about a gallon, and refill these from the jerrycans on a daily basis.

The living space is separated from the cab, and accessed through a weathered stable door at the rear of the wagon. On entering, a wooden dresser sits immediately to your left, holding a miscellany of objects and belongings. Front and centre is my heavy enamel saucepan, capable of holding two gallons, an indispensable piece of equipment, especially during the cold, dark winter months when it holds a good broth or stew that takes just minutes to warm through. There is also room on the dresser for a copper-bodied oil lamp with

AUGUST

a sizeable reservoir containing enough paraffin to last several evenings, even during the longest nights of December and January. Apart from candles, it is the only source of lighting in here. Next to this sits an antique biscuit tin that houses a small first-aid kit. The bandages, plasters and eyewash it contains are, I believe, out of date, though I suspect they would still be sufficient to deal with any minor injury.

Amongst the china plates and cups that adorn the dresser shelves are various ornaments that I have collected since living here. A modest selection of antique clay tobacco pipes found at different locations whilst hedging; Yorkshire and Suffolk pattern billhooks hanging from handles worn smooth by the men who used them in decades past; a collection of Victorian glass bottles and tea tins, alongside several antiquarian books written by countrymen from the last two centuries. I have a particular fondness for older books and have found much wisdom within the pages of my modest collection. A particular favourite is the author Richard Jefferies, who was raised in rural Wiltshire in the nineteenth century and produced several works about the British countryside at that time. On occasion, his writing has seemed almost to speak to me directly. 'For the spirit of nature stays and will always be here, no matter how high a pinnacle of thought the human mind will attain; still the sweet air, and the hills, and the sun, will always be with us.'

Opposite the dresser stands a gas cooker that includes an oven and grill as well as four gas rings, which is more than adequate for my cooking needs. Next to this, a handmade oak kitchen unit houses a ceramic Belfast sink that can hold several gallons of water, which I rely on for bathing, as well as washing up. The larder below the sink contains an array of tinned meats, soups and powdered milk; there is no refrigerator in here and perishable products must be stowed outside or in the cab of the lorry during the cooler months to prevent them from spoiling.

A cast-iron woodstove sits next to the oak unit and although it is relatively small by most standards, it is nevertheless capable of warming the interior of the wagon within minutes. I collect wood from the hedgerow as I work and dry it on the slate hearth around the stove, the curing wood filling the living space with a pleasing organic aroma that mixes with the more synthetic scent of burning paraffin given off by the oil lamp. I try to keep a bowl of oranges in here throughout the winter months, drying their peel on the hotplate of the stove, which not only provides an excellent accelerant for the fire but also fills the living space with a sweet, citrus scent. A well-worn armchair is positioned conveniently in front of the fire so that I can warm myself through thoroughly at the end of winter days spent outside in the worst of the weather. Then I have an antiquated, green-painted footstool to prop my feet on and warm them by the flames, reclining in the chair and sitting in complete comfort.

The bed is above the cab of the lorry, enclosed on all sides and accessed by three oak steps positioned at the head

AUGUST

end of the sleeping compartment. A small window sits just above the pillows and when opened provides ventilation both on hot summer evenings and during the long winter nights when the rising heat from the stove can make the compact sleeping area feel stifling.

I have always been drawn to this life amongst the woodlands and hedgerows. For me, the lure of modern living holds little appeal and it has always been the old country ways, their knowledge and wisdom, that have inspired me. Since starting out as an apprentice woodsman in an isolated Dorset woodland nearly thirty years ago, I have turned my hand to many rural practices, listening willingly to advice and guidance passed down from men whose years and experience exceed mine by several decades. Of the varied skills practised by the woodsman, it is the traditional management technique of hedgelaying that has become my favoured employment.

Hedgerows have been a feature of the British landscape since the very onset of agriculture, when our Bronze Age ancestors clearing land for farming first spared strips of the ancient wildwood, leaving trees and woody shrubs in linear patterns to act as boundary markers and serve as fences for containing animals. From this time onwards hedgerows have been a consistent part of British agriculture, though it wasn't until the thirteenth century and the beginnings of the enclosure movement that the planting of hedgerows really began to proliferate. As the expanding cloth trade and increasing demand for wool saw a huge growth in sheep farming, new pastures were enclosed to secure the multiplying flocks and mile upon mile of new hedgerow was planted.

Initially, labourers sourced plants from scrub and woodland for hedge planting, and a wide variety of species were used according to local availability, but as the momentum of enclosure increased, it became necessary to establish nurseries across the countryside to provide the plants needed, and hawthorn soon became the preferred plant for these new boundaries. The hawthorn was fast growing, tolerant of shade and both hardy and resilient. Planted in a dense row, the spiny, horizontal branches knitted together into a tight web of thorns that formed a living barrier capable of deterring even the most determined of livestock.

The enclosure movement continued into the early twentieth century and over its course, millions of acres of pasture were eventually divided by many thousands of miles of hedgerows. The lush green lowlands of Britain were gradually transformed over this period to eventually become the instantly recognisable patchwork of small fields and meadows for which our countryside is known, a land enclosed by thorns and briars.

These field boundaries, however, are more than just living fences. Steeped in our political, cultural and agricultural past,

AUGUST

they have come to define much of the British countryside – and to preserve many of its best-loved inhabitants. A great deal of our farmland wildlife, the songbirds and invertebrates, as well as smaller mammals, have come to rely on these arboreal arteries that link much of our remaining woodlands and enable safe passage across open pasture. They provide food, shelter and nesting sites for many threatened species.

The hedgerow left to itself, however, will decline and eventually fail. A hedgerow that is maintained correctly and that has both periodic laying as well as trimming incorporated into its management will provide the most beneficial habitat, and will go on to do so for many centuries. This skill of hedgelaying was once widespread throughout much of England and Wales, and by the seventeenth century the 'hedger' was a familiar figure in the English countryside. For many centuries the maintenance of the countryside's hedgerows was an essential part of the farming calendar and the laying process saw that these boundaries remained dense and impenetrable to the livestock they contained, as well as preserving them as one of Britain's premier habitats for nesting birds and other farmland wildlife. In recent decades, however, as agriculture has become increasingly intensified, the hedger has dwindled into a rare figure, and with his departure it seems that the linear woodlands he worked have become little more than an agricultural bygone. Impractical on the modern, intensively managed farms, the ongoing maintenance of hedgerows has increasingly been seen as a burden, and the route they once carved across fields and pastures has become a hindrance to the ever-growing range of farm machinery. In the second half of the last century, half of all British hedgerows were grubbed out.

The demise of hedgerows, combined with the improper maintenance of many of those that remain, is now reflected directly in the decline of our rural species as their habitats are destroyed or degraded. As we lose the old wisdom and the ancient skills that see the countryside managed regeneratively, the land has undoubtedly suffered.

But it is not all bad news. Despite the loss of well over one hundred thousand miles of hedgerows since the middle of the last century, they have not been lost altogether. Here in the West Country they remain a prominent feature of the farmland, and are even protected by law. And although today managing hedges by laying is the exception rather than the rule, interest in their preservation is expanding and the skill remains in demand. Indeed, as August draws to a close and I look to the hedging season ahead, my calendar is full.

It is during the shortening days of the autumn months, when the September mists return and the morning dew settles on the pastures once more, that the hedger begins his work. As the light begins to retreat in the later months

AUGUST

of the year and the ground is cooled by first rain and then frost, the plants of the hedgerow begin to fall dormant, drawing the nutrient-rich sap into their roots before eventually casting their leaves to the November gales. Despite the adverse weather conditions, this is the best time of year for laying.

A hedgerow should be laid every fifteen or twenty years to see it maintained in a prime condition, and a good hedgelayer should be able to achieve a 'chain' over a day's work, a length of hedgerow equal to sixty-six feet. The length to be worked on must first be cleared of any dead or dying wood, brambles and briars must be cut away and removed to leave only the stems that will be used in the creation of the barrier. A first observation of hedgelaying by somebody ignorant of the technique, can seem severe and may appear to be a destructive rather than a regenerative practice. But the plants that make up the hedgerows have evolved to tolerate and thrive on this adversity. When one stem is lost or damaged, for example by browsing animals, the plant's response is to send up vigorous new growth, with several new stems taking its place. The hedger mimics this attention by carefully cutting and pruning to encourage a structure that will regenerate over time and continue to grow.

Clearing all excess material, the hedger will retain a line of clean and straight stems, the best that are available in the row, pruned and freed from any entangling briars that will readily find their way into the hedge, the seeds brought in by the birds and animals that make their homes there. Using a razor-sharp cutting tool, the hedger slices into the back of the first stem, cutting downward at a steep angle from a foot or so above the ground to perhaps four fifths of the way through, being careful to retain a tongue of living wood at the stem's front edge. Although bark and heartwood have been severed, the retention of the malleable strip of wood ensures that sap is still able to rise in the spring and that the stem will continue to live. This practice is known as pleaching and once the cut has been made, the hedger refers to the stem as a 'pleacher'. With the pleaching cut made, the stem is now carefully laid over until it rests diagonally in the row, still anchored to the stump by the tongue of wood. The stump must now be dressed tidily, as it is from here that the majority of regrowth will come. The upright spike of wood that remains on the stump after the pleaching cut, known as the heel, is removed to ensure that water and leaf mould do not collect there, as these could invite rot into the plant. This process is repeated with each stem in turn, the pleachers resting against one another tidily in their diagonal position, like plates in a rack.

Once all the stems have been pleached and laid into position the structure is reinforced by a line of stakes along its extent. Generally, these are stout lengths of hazel measuring about five foot six which have been cut specifically for this purpose. Each stake is pointed at its thickest end

AUGUST

which is driven firmly into the ground along the row, the spacing between them traditionally measured by the distance between the hedger's elbow and fingertips. By this time the structure is beginning to resemble a fence, with a spine of stakes running through the newly pleached plants. Working now in the opposite direction to the pleachers, the structure is tightly bound to increase rigidity further still. The binding is done by weaving smooth hazel rods, cut at about ten or twelve feet in length, between the stakes. By adding a new rod into the weave at every stake the whole construct is locked firmly together, and the woven hazel not only provides an attractive spiralling pattern but also creates a continuous wooden rope along the length of the newly laid hedgerow.

The final task is to trim the stakes to the height of a clenched fist above the binding, for a tidy finish. Generally this is done to leave a point which allows water to run off easily as well as providing an extra deterrent to livestock. While the finished hedgerow may look sparse on completion, over the course of its first spring the rising sap will stimulate growth and see the boundary increase still further in density and strength.

Great pride is taken in a well-laid hedge by landowners and hedgers alike, and each craftsman has their own particular style of working. Just as an unsigned painting or sculpture may be recognisable by style alone, so the work of individual hedgers may be recognised by their individual preferences and intricacies. Hedgelaying is both sculptural and functional and when carried out correctly, is a work of art in itself.

I had been staying in the busy yard of a West Dorset dairy for almost six months and the harvest, the culmination of the year's work, was drawing to a close. Standing by the open stable door of the wagon and looking out over the field, I noticed countless threads of spider's silk clinging softly to the freshly cut corn stubble, the late-summer breeze causing them to ripple gently and catch the evening sunlight. In the wagon behind me, the kettle sat on the gas hob, grumbling away but apparently in no hurry to come to the boil. Eventually steam began to gush from the soot-blackened spout and when I returned to my spot at the door it was with a cup of hot, sweet tea. My attention was caught by a pair of goldfinches that streaked past the open door before settling momentarily amongst the thistles that grew on the headland just outside the wagon. By this time of the year the thistles were wilted and dry and tufts of the thistledown were lifted by the breeze as the finches took flight, only to collect in clumps on the teasles, hogweed and elder that grew behind the old cattle shed close to my pitch.

I had been there all summer, parked next to the decaying outbuilding. Several swallows' nests, now empty, nestled tightly where the overslung corrugated-tin sheets met the

AUGUST

building's wooden joists. The swallows still darted around the farmyard and across the recently harvested corn and barley fields, catching insects on the wing and calling in their distinctive chatter all the while. Sparrows flitted between the gaps in the Yorkshire boarding which made up half the walls of the building. A thick stem of ivy had climbed the cast-iron downpipe and spread to smother much of the back wall. Upon my arrival six months earlier, I had noticed a robin who had built her nest amongst the creeper's waxy leaves; the cover they provided had proved to be an excellent choice for her nest and she had successfully fledged her brood less than a month later.

As I stood in the doorway of the wagon the evening began to feel autumnal. It wasn't cold, but there was definitely a change in the air. There was a shift in the quality of the light, and as the sun descended ever lower in the west, it cast long shadows across the antiquated pine desk that sat between the dresser and armchair. When I looked out at the yard, the end of summer was apparent. The rectangular straw bales that sat neatly at intervals around the field would soon be loaded on to the farm's trailers and brought down to the shed where they would be used for bedding up the cattle over the winter. We had been enjoying a prolonged dry spell and with the settled weather forecast to continue, a good deal of urgency had been absent from the harvest. The bales could wait for now, and by the time they were brought in over the coming days, I would be gone from this place. After six months here I was looking forward to the move. Although by late winter I am weary of the thorns and bramble, by the end of August I am equally enthusiastic to get back amongst them and continue my work.

It was the final day of August when I pulled away from the dairy in West Dorset. The wagon was packed up, the ornaments and contents of the dresser stowed safely in a wooden trunk beneath the desk, as I navigated my way carefully between the cattle sheds and along the farm's compacted chalk track before eventually turning out into the lane. The road took me eastwards, passing a row of old brick-and-flint cottages typical of this part of the county, where the flints cast up during the cultivation of the land have always been collected and incorporated into many of the local buildings. These cottages would once have served as dwellings for the labourers and their families who worked on the farm I had just departed from. Now, however, they were owned by city commuters who made their living elsewhere. I crossed an ancient, narrow stone bridge that spanned the gin-clear waters of the chalk stream that ran through the valley before turning north, passing through the Georgian market town of Blandford Forum, crossing Dorset's River Stour and following the route onwards, across the Blackmore Vale which lay to my west.

I made my way through the picturesque Dorset villages north of Blandford, admiring the quaint cottages and craftsmanship that was evident on their thatched roofs. From the

AUGUST

straw ridges of the thatch, among the hazel latticework that held the ridge firm, were figures of different animals, each created from knotted thatching reed. The various figures have long been used by different thatching contractors as a means of distinguishing between their work, and much of the local wildlife had been represented. On one, a fox chased a hare along the ridge, while on another a cock pheasant sat proudly at the gable end. Such craftsmanship was clearly still prized here.

As the villages gave way to open farmland, I made the gradual climb towards the hilltop town of Shaftesbury where I was able to see over the tops of the hedgerows and out across the vale. With the prolonged spell of dry and settled weather the harvest had already finished on many of the farms here, although towards the horizon I could see the distinctive dust cloud stirred up by a combine harvester still working on one of the bigger estates in the distance. The countryside looked parched, a lifeless beige after one of the driest summers in some years, and amongst the recently cropped fields of corn and scorched pastures, it was only the pockets of woodland and the hedgerows connecting them that remained lush and green. They acted as an oasis to the farmland wildlife which relied on them, especially in these conditions.

I travelled on through the dusty landscape, and after little more than an hour I arrived at my destination, Eweleaze Farm, to begin my first hedging contract of the season.

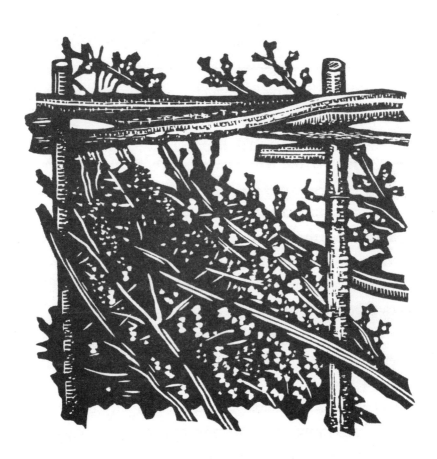

2

September

My arrival at Eweleaze Farm was obviously anticipated. A woman clad in jodhpurs and riding boots left her work in the stableyard to greet me and direct the wagon past a series of outbuildings and on to a patch of hard ground next to a wooden sheep shed. The farm, as its name suggests, keeps a modest flock of Dorset sheep which would come to the shed for lambing, shearing and to shelter from the worst of the winter weather. A tap on the side of the building would provide me with water for the duration of my stay and a lavatory was located in one of the outbuildings just a short walk down the track. There was a completely different feel to this pitch. It was quiet here. Throughout the summer I had become used to the busy yard of a commercial dairy, and although I had enjoyed the company of the labourers and never felt unwelcome, I was pleased to be away from the noise and commotion of the cattle sheds and farmyard.

It didn't take long to get the wagon's living space organised and by the time the kettle had boiled on the gas hob, the majority of my belongings were in order. I sat on the steps of the wagon with my tea, watching swallows dancing around an old ash tree in the hedgerow in the half-light of the late-summer evening. As I looked across the broad pasture to a wooded hillside in the distance, I was reminded how much I had missed these quiet corners of the countryside.

Eweleaze is an old Duchy farm, previously let by the estate owned by the Duchy of Cornwall. The farm had been a relatively large dairy unit for many years before the previous owners had sold the cows and moved the operation over to arable. The new owners told me that the farm had been on a downward cycle for some time before the sale, the pastures heavily overgrazed and barely able to sustain livestock, and that although the initial move to arable had been promising, yielding several years of plentiful harvest, the continued deep ploughing of the land eventually saw hundreds of years of topsoil lost and the deeper clay brought to the surface. In the years running up to the sale, the crops had begun to fail. Other areas of the farm had also fared badly. Alder and willow had grown into dense thickets all along the stream that marked the farm's boundary. Unchecked bramble growth was smothering the banks and climbing ever higher into the trees, casting continuous shade on the water which was itself depleted of much of the life it would once have sustained. The hedgerows had been trimmed annually, but in the absence of any laying being incorporated into their management, they had thinned and declined considerably. Stunted and twisted blackthorn, hawthorn and spindle were now giving way to the more resilient bramble, and the resulting gaps had been filled crudely with rusting metal hurdles and a tangle of barbed wire. The new owners had purchased the farm from the estate around a decade ago and embarked on an extensive restoration project that they hoped would eventually see a return of the rich biodiversity that would once have been abundant on the farm.

The contract here was a lengthy one: eight hundred metres of primarily blackthorn hedgerow that was to be laid over two successive winter seasons. Blackthorn makes a favourable hedgerow; the sharp spines are an excellent deterrent to livestock whilst also providing refuge for songbirds and other wildlife. It is, however, notoriously difficult to work with. Although the stems pleach well, blackthorn tends to grow in distorted shapes, tangling itself amongst other plants in the hedgerow. This of course makes an excellent barrier once laid into place, but requires firm persuasion to bring it into line. A good pair of gloves offers some protection, although once wet even the toughest leather becomes susceptible to penetration, and even when the gloves are dry the tough thorns will pass through the lighter seams. Splinters are inevitable, and the needle-like barbs of the blackthorn will embed deeply in a hand or arm, irritating the skin and joints. It is to be expected when working among the blackthorn at this time of year to have swollen and sore fingers from one or more spines deep in a knuckle, or worse, under a fingernail. My first-aid box, although perhaps somewhat underequipped, always contains several sharp sewing needles for extracting the spines from tender hands. I remove these splinters as quickly as possible, as if left they can fester, causing severe swelling and discomfort, and in the worst cases leading to infection and a course of antibiotics. It is wise then to be mindful when working with blackthorn, but go carefully and a most satisfactory hedgerow can be built.

The hedgerows to be laid at Eweleaze were wide, although not particularly tall. Blackthorn is constantly on the move, sending vigorous suckering growth up from its roots. If left unchecked the hedgerow will gradually creep into the pasture, gaining several yards each growing season. Left completely to its own devices it will eventually form dense thickets of scrub or will grow tall and leggy, the thorns open enough to allow sheep to stray into the hedgerow's base where they can become entangled and perish. While blackthorn scrub is a valuable habitat, a carefully crafted hedgerow will provide all the benefits of this habitat, at the same time also acting as a living fence.

The traditional tools for hedging work are the billhook and axe, and I have several of each. The majority of hedgerows that contractors face today, however, have often been left for many decades. More often than not they will contain semi-mature trees that must be coppiced or pollarded to reinstate the hedge line, and predictably the contract here at Eweleaze contained several of these. Clumps of multi-stemmed ash that would once have been maintained as part of the hedgerow now stood at forty or fifty feet high, only succeeding in shading out the thorns below. The long straight trunks and relatively open canopies provided little in terms of either habitat or aesthetic value and so would be felled and coppiced back to ground level, the wood split into logs to warm the farmhouse. Perhaps one or two of the finer specimens could be allowed to grow on, which would then fill out with the competition from their neighbours removed, while the cut stumps from the felled trees would regenerate and send up fresh growth. In this way, coppicing keeps a tree in a juvenile state, extending its life cycle long past its natural span. As the new growth comes up alongside that of the blackthorn and other hedgerow shrubs it can then be maintained as part of the rejuvenating hedge line by trimming. When the time comes to lay the hedgerow again, the tree will be left to grow on until it reaches a suitable size to pleach and build into the hedgerow. Although possible with hand tools, this kind of heavy work is more effectively carried out with a chainsaw and I know of no professional contractors who use only the traditional billhook and axe.

The thorns at Eweleaze were to be laid in the Midland style. Of all the styles, of which there are perhaps thirty or more still regularly practised, in my opinion the Midland is not only the most robust from the outset, but also the most attractive. While all rely on pleaching and laying to reinvigorate and increase density within the hedgerow, over time different styles of hedging developed across the different counties and regions of England and Wales, varying according to the agricultural practices of each district and the materials available. The styles can essentially be divided into two main categories: those that are supported by stakes and are laid upright to form a barrier from the outset; and those that are laid horizontally, the pleached stems in close contact with the ground where they can then regenerate to form a dense hedgerow in the coming seasons. Here in the southwest, the hedgerows are generally laid in the latter style, but are usually set upon a steep bank which forms the main barrier to livestock. However, over decades of neglect many of the banks here at Eweleaze had crumbled, and now formed no barrier at all. In cases such as this, where the boundary has long been in decline and needs thorough restoration, laying in one of the upright styles, such as the

SEPTEMBER

Midland that I would be using, provides an immediately effective barrier despite the lack of bank.

The upright styles can be divided into a further two categories: single- and double-brushed. If a hedgerow is to have livestock kept on both sides, it needs to be double-brushed, meaning that the bushy bulk of the laid stems falls to either side of the central row of stakes, laid to the left and right alternately. An example of this is the South of England style which is commonly seen across Hampshire, Surrey and Berkshire and the surrounding counties, where its main function is the containment of sheep.

The Midland hedge, however, is a single-brushed hedge and its primary function is the containment of cattle; for this reason it is often referred to as the Midland Bullock Hedge. The Midland style has an attractive nearside face that clearly shows the pleaching cuts below the woven binding; it is from this side that the regrowth will come and so this faces out, away from the side where livestock will be kept. The stakes then run just adjacent to the line of stumps, rather than down the centre of the hedgerow, and all of the brushwood from the pleached stems is directed into the far, or field, side. This ensures that any livestock approaching the boundary are confronted with a dense wall of thorns, while the tightly bound stakes hold the structure firm. The staking and binding will rot over several seasons, but by the time this happens, they will have served their purpose and the regrowth would have reinforced the boundary sufficiently.

It has been suggested that in the past a rural native lost in the countryside could gather their rough whereabouts by the style of hedging they encountered. In this day and age, recent agricultural practices in some districts may mean they are lucky to encounter a hedgerow at all.

I was out of bed by half-past five in the morning, keen to make the most of the light that September still affords. During these first daylight hours it's noticeably quiet and is, in my opinion, the best time of the day. Most mornings now saw a heavy dew and there was a vale of mist in the hollow leading down to the river just a stone's throw from my pitch, beyond the alder and willow scrub. Although the morning air was fresh, I was quite comfortable standing by the open stable door whilst waiting for the kettle to boil, and once the teapot had been filled, I poured the remaining water into the Belfast sink. By the time I'd drunk the first of several cups of tea, the ceramic bowl had cooled the water sufficiently to allow me to run a hot flannel over my face and neck.

It took little more than ten minutes to walk from the wagon to the hedgerow. An old canvas satchel slung over my shoulder contained everything I would need for the day: leather gloves, a selection of chainsaw files and a spanner, alongside a flask of tea, a small loaf of bread and a jar of

SEPTEMBER

peanut butter, several pieces of fruit and some dark chocolate.

Halfway along the hedgerow's length was a wooden field gate, nestled beneath one of three mature oaks which rose from the entanglement of thorns. It sat conveniently at the highest point along the row and I decided to start my work here, working away from the gate in each direction, so that the hedges on either side were laid towards each other and the stems all faced uphill. Standard practice dictates that a hedge should always be laid facing uphill, and if it sits on level ground then it should be laid in the direction of the rising sun. Although the undulation in the pasture here was subtle, there was no doubt that the ground sloped away gently from either side of the gateway.

Placing my satchel at the base of the oak, I paced out twenty metres along the hedge line and drove a hazel stake into the ground to mark the point I needed to reach by the end of my day's work. If I made good time I could work on beyond the marker, but I needed to achieve at least this distance each day to ensure the contract was completed in time. The season ahead was busy and all the work needed to be completed by spring. The old country wisdom states that a hedge can be laid during any month with an 'R' in it. Historically, hedging generally started in September when the harvest had come to a close and the labourers could be put to work on the estate boundaries, and was completed by the following April, when the sap was rising and the plants were once more coming into leaf. In recent years, April has felt late to be finishing my work; the birds are already stirring by late February and by March they begin nesting. By the second week of March I leave the hedgerows to the birds. I have heard of hedgers leaving a reward in the hedge, a bottle of cider or a meat pasty perhaps, a carrot dangled to encourage them forwards so the work is finished in time.

After casting an eye across the first length to be worked on, I pulled on my gloves and set about clearing the old fencing that ran throughout the hedgerow. Barbed wire had been used to discourage livestock from straying into the scrub and there were now three or more old fence lines in various stages of decay to contend with. The oldest, which was by now barely visible, was easily snapped by a gloved hand and gave a clue to the original width of the hedge. There were old pleachers in here too, horizontal stems now almost the width of a gate post which revealed that this hedge, like most, had once been maintained by laying, almost certainly in the first half of the last century, likely between the wars. I used a pair of heavy-duty fencing pliers to cut the wire, and drew the rusting strands steadily from the thorn. Once removed, old wire must be coiled neatly and piled in a convenient gateway or at the base of a tree where it won't be forgotten, so it can be collected and disposed of by the farmer. It is important to make sure that rotting posts are collected too, as they are likely to contain staples

SEPTEMBER

and nails that could cause serious problems if caught up by machinery or grazing livestock.

With the wire removed I could begin cutting away the scrubby, suckering growth with the chainsaw. The hedge was perhaps fifteen or twenty feet wide in places, and needed to be reduced to a corridor of no more than four or five feet, which would then need further thinning out to leave only the stems that were to be pleached to form the new hedgerow.

The initial clearance work took most of the morning and it felt like hard going in the heat of that early September. I worked my way steadily along the length to be completed, taking a few feet of scrub at a time, cutting it as close to the ground as I was able. Despite the increasing warmth of the day and the physical nature of the work I made sure to keep the sleeves of my shirt rolled down and the cuffs buttoned tightly, but even so the blackthorn and brambles snatched at my clothes and gloves, and more than once a twisted stem caught the mesh visor or ear muffs of the helmet I wear when using the chainsaw and toppled it from my head. The bushy tops of the thorns were intertwined with each other and tied up still further with dog rose, brambles and honeysuckle. I had to put the chainsaw aside at times, grasp the entangled stems with both hands and wrestle them free.

I dragged the cleared brushwood several yards into the field before stacking it parallel to the hedgerow in a continuous drift. I'm mindful of making as tidy a job as possible during the initial clearance of a hedge line, ensuring a straightforward task for the farmer when it's time to 'burn up'. In the past it was the hedger who would carry out the burning of the excess material, collecting the waste into piles along the hedgerow and then burning them in the field as he went, the resulting ash from the fires improving the soil. In more recent times, however, this work has been carried out by the farmer with the use of a tractor. The brushwood can be collected with the aid of a hydraulic grab and taken to a convenient location on the farm where a bonfire can be built, although the ash is generally still ploughed in to improve the soil.

I always keep my eyes open for a good walking stick when cutting blackthorn. Often, amongst the scrub a suitably straight length of thorn, say four or five feet, with a fork or appealing shape of some kind, will catch my eye. I put these to one side and take them home at the end of the day where they can be left to season over the next twelve months or so. Blackthorn makes the best walking sticks. Once the thorns have been carefully removed with a sharp pocket knife and any mould or dirt cleaned from the shank, a light coating of oil will reveal a beautiful deep mauve and mottled crimson colouration that contrasts exceptionally well with the lighter, buttery yellow of the heartwood that is revealed in the shaping of the handle. Alternatively, an appropriately shaped piece of deer antler can be fashioned into a handle; a piece of sika antler with

SEPTEMBER

a collar of silver, or a roe antler, which flares conveniently at the base so that when shank and antler are fixed together the joint is hidden. A thin wrap of lead and a rubber ferrule added to the base of the shank, coupled with a leather lanyard where the antler and shank join, make a particularly attractive wading stick for the salmon or trout fisherman.

It is at this stage of the work that I also put aside any deadwood that I come across: ash or oak branches that have fallen from the hedgerow trees and have been cradled in the thorns, kept from rotting on the damp earth below, or standing dead thorn wood that has been propped up by its neighbours on either side. These are cut into the small logs needed for my woodstove in the wagon or for an outside fire in the evenings. I keep an old feed sack folded up in my satchel and once filled to the top it carries enough wood to heat the wagon all night.

By the time the clearance work was complete it was a little after midday and I was grateful for the shade provided by the oak tree next to the gateway. The heat of the summer can linger for much of September, and despite the noticeably retreating light and cooler evenings, in the middle of the day it felt like July. The shade of the oak and the warm breeze blowing across my shirt were most welcome, though the oak boughs were already heavy with acorns, and as I poured the tea from my flask several fell on the ground around me. I pulled my old tweed cap down tightly to my ears and took my chances with the falling acorns, sitting there long enough to consume a thick crust of bread primed with a good measure of peanut butter, and two cups of strong tea. Somewhat revived, I was now keen to press on. A longer break only makes it harder to get back to the afternoon's work; instead, I keep in mind the comfort of the old armchair and the pleasure of September evenings spent drinking tea by the open door in the old wagon. The sooner I could achieve my day's objective the sooner I could settle for the evening.

The hedge by now looked less daunting, a double row of blackthorn no broader than five feet and interspersed with several hawthorns and spindles along its length. There were other species, too, peppered amongst the thorn: guelder rose, field maple, holly and young oak, elder, hazel and ash. They accompanied the ever-present bramble and dog rose, honeysuckle and ivy which had migrated here from surrounding woodland, carried by song thrush and wood pigeon, fox or badger, to grow amongst the blackthorn from which the hedgerow was originally constructed centuries before.

It was time to start pleaching. I had removed any lower branches that protruded from the thorn trees, so that I had clean stems to work with. Although the majority of regeneration will rise from the stump, new growth will come

SEPTEMBER

from any of the cuts I have made, and tidying each stem before pleaching not only makes it easier to work along each length but also encourages regeneration along the diagonally laid stem to increase density still further. Nature will continue to tend and improve the structure once my work is finished. The bushier tops of the thorns to be pleached had been mostly untangled and without the scrub to support them some now hung limply. I had carefully selected the best available, and they pleached easily, the sharp teeth of the chainsaw making the cut effortlessly down to the base of the plant. I'm always careful to avoid bringing the chain in contact with the earth and flint at the base of a hedge, as this would render it useless. A dull chain not only slows the work considerably but is also dangerous, as the user must force the cutting blade through the wood, making accidents more likely. Alone in a hedgerow, often a mile or more from a surfaced road, is not a good place to be dealing with a chainsaw injury. The chain should pass through the wood as a hot knife through butter.

 The stems were pleached in turn. After each cut, I placed the saw on the ground and carefully laid each spiney stem against the previous one, lowering them gently into position, using a convenient fork or kink in their form to loosely lock them together. In this way, I methodically worked my way along the row until all the stems had been pleached, the contorted branches and thorns biting into one another and holding the structure together. It is amongst these spiny prunus that the farmland songbirds will nest next spring, raising their broods away from the watchful eye of the magpie and jay who patrol the hedgerows and make short work of nests in more open, derelict hedgerows.

By mid-afternoon it was a relief to put the saw down and enjoy the last of the tea from my flask, now tepid and stewed but welcome none the less, before moving on to staking and binding, my final tasks for the day. I put the stakes in position along the pleachers, the smooth silvery hazel easily penetrating the pleached thorns. The ground at the base of the hedge was hard and strewn with ancient roots, but the stakes were sharp, and once I was happy with their position I drove them in firmly with a sledgehammer, readjusting each stake if needed to ensure the line remained straight. I believe it is worth taking the time and care to concentrate on these details. When laid properly, a finished hedgerow speaks of the craft that was once so prevalent in the countryside but has now been almost forgotten, and shows a high regard not only for traditional skills but also for the environment and its sympathetic management.

The binding of the hedgerow is my favourite job. After the exertion of the clearance work that morning and the noise of the saw, I finished the day's work with a quiet hour weaving the long hazel rods between the stakes. I started with four rods and placed two on each side of the first stake before twisting each pair together and weaving them alternately, adding a fresh binding rod at each stake and repeating the process throughout the newly laid length. The tension in each rod was considerable and it required some effort to pass them between the stakes. I always try to use any irregularity in the rods to my advantage, using thicker rods to bring wayward stakes back into line and thinner ones where the stakes are not held so firmly in the earth. In this way, I maintain the consistency of the line.

The binding gripped the stakes so firmly that I had to use a lump hammer to knock them down to the desired height, stepping back often to ensure the height remained consistent and evenly followed the contour of the ground. As the binding was knocked into place, it compressed the unruly thorns into a tight mass, the uniformity and straight lines of the stakes and binding contrasting with the thorn that, although laid tidily, was still wild and natural in its form. Finally, I used a pruning saw to cut the tops from each stake, leaving a tidy line. I angled the points subtly so that the cut face of each point matched the pleaching cuts of the laid stems. When looking down the hedgerow in the direction of the laying, the work and time that had gone into its construction were clearly evident.

It was five o'clock by the time the binding was on and that day's section complete. I pulled the marker stake out of the ground and paced out the twenty metres for tomorrow's stretch, casting an eye over the work that would need to be done. It looked thinner, likely an easier day and earlier finish.

The air was beginning to cool as I headed back to the wagon. Stepping over the stile that led into home field on the farm, I stopped to pick some blackberries from the hedge. A buzzard called overhead, and looking up I saw swallows catching insects on the wing and chattering incessantly. Next month these birds would be gone, making their long migration south, through Spain and across the Strait of Gibraltar to the African coast and the Sahara beyond. I, too, would be on the move.

3

October

Although I am rarely cold in the wagon, a distinct chill woke me just before dawn. Gradually, the subtle autumnal light brightened the living space and the robin's song roused me. After descending from the sleeping compartment I lit the blackened and frayed wick of the oil lamp before hurrying to pull on the chilled vest and shirt that lay draped untidily across the old armchair. The bitterly cold morning warranted the lighting of the woodstove, but I refrained, knowing the flames from the stove would encourage me to stay in the confines of the wagon, and do little to aid my productivity. I wanted to get moving, if only to keep warm. After washing my face and pulling on my cap, I descended the steps with my tea. My hands now carried cuts and splinters from working amongst the thorns for more than a month and I clasped them gratefully around the comforting heat of the mug. There was a light frost on the grass, which was still stunted and brown from the summer drought. The frost was the first of the season and I took some moments whilst drinking my tea to enjoy the promise of this crisp, new day. The early mist only partly obscured the alder and willow scrub that grew here and amongst it I noticed the golden leaves of a lone, self-seeded field maple, wrestling for light amongst the other vegetation. It had grown twisted and contorted in its efforts, but now, in full autumnal colour, it outshone its neighbours. As I sipped my tea, the sun rose above the horizon in the east, a shimmering crimson orb; the mist allowed me to stare directly at it, shielding its glare without diminishing its beauty. On my returning to the wagon, the gas hob, which I had left alight, had eased the chill, although my breath was still clearly visible. I prepared eggs for breakfast, then filled my flask and satchel, before making my way to the hedgerow. These October days were often still pleasantly warm, but this beautiful morning whispered of the winter to come.

By the middle of the month I had completed a good deal of the Eweleaze hedgerows. Initially I had worked eastward from the gateway, and having completed that section, I had now started work from the other side of the gate, working in a westerly direction back towards my pitch at the sheep shed. As the month progressed the hedgerow's autumnal colours were becoming ever more vivid – vermilion red, copper and gold. This seasonal display can vary greatly in its vibrancy from year to year, and this was amongst the best I was able to recall, perhaps due to the summer drought that had put the trees and shrubs of the hedgerow under some stress. Then I reminded myself how often I had thought this, and wondered why. Was it the brevity of these exceptional displays that makes each seem more spectacular than the last?

With the uniformly green foliage of the summer now faded, the progressing season made the different species that composed the boundary more conspicuous. The leaves of the blackthorn had turned a beautiful deep russet. The delicate leaves of the hawthorns had faded to a pale yellow and the red haw berries that were yet to be taken by the birds decorated the trees' matt grey stems. The glassy, translucent berries of the guelder rose hung in heavy clusters amongst the plant's reddening foliage, its maple-like leaf still retaining hints of fading green, although most were turning yellow or darkening to blood red. I came across this shrub often, especially in the hedgerows of Dorset and Wiltshire where it thrived in the chalky soil. The spindle tree, too, had gained a foothold here on the chalk and its distinctive, ripening pink fruit was now splitting to reveal the orange, corn-like seeds within. The holly remained the dark, glossy evergreen it would be throughout the year but was also now adorned with berries.

I had pleached and laid most of the holly into the hedgerow, though I had left several to grow on, selecting those which had a particularly pleasing form. In past times, holly trees were often placed at regular intervals along a hedgerow's length. The tree's distinct form was easily made out by the ploughman when working across the fields towards it, and an occasional glance up while treading the furrow would ensure he remained on a steady course. Its presence in the hedgerow is now only a fading memory of the purpose it once served, but the holly, and any tree that is left to reach maturity, still has an important role to play. Rising clear of the boundary they provide a song post for birds making a bid on nesting territories in the lengthening days of late winter, and are used by bats to navigate the meadows on summer evenings as they embark on their night-time forays in search of moths and the other invertebrates on which they feed.

Of all the shrubs that make up the hedgerow, in my opinion it is the field maple that is the most striking during this month. By the later days of October, its delicately lobed leaf has been gilded by the cooling earth and autumnal mists. Like the guelder rose and spindle, the field maple thrives in the chalk earth and is well suited as a hedgerow tree so, as with the holly, I left several of the maples where I came across them. While the shrubs of the hedgerow are tolerant of shade to a degree, when under the permanent shadow cast by large deciduous trees in full leaf they become increasingly stunted. As the tree reaches its full potential the hedgerow shrubs beneath thin and die back, receding to the peripheries of the shade cast by the expanding canopy. Oak, beech, sycamore and others, although undoubtedly assets to our farmland that are to be cherished and conserved, will still often put any hedgerow growing beneath them on the back foot. The field maple left to grow to maturity, however, will rarely cause this problem due to its more subtle nature and compact form. Despite becoming a tree of not inconsiderable proportions, it will never equal oak, beech or even its cousin the sycamore in size. A field maple allowed to grow on will share the light amicably with the other plants of the hedgerow, providing the benefits of its additional habitat while still allowing the plants below to thrive. This tree is the only native maple on these isles and its attributes have seen it utilised amongst our hedgerows for centuries. Whether the field maple is laid and worked into the boundary or given free rein to mature naturally, its presence amongst the thorns is always most welcome.

As the autumn reached its peak, it was only the oak trees that retained the majority of green in their leaves, although by now these too were looking ragged. As their canopies caught the October breeze the leaves sounded crisp and brittle and the lucid greens of May, of vibrant oak crowns saturated by rising sap, felt a distant memory.

The autumnal vibrancy of the past few weeks was fading.

The thorns began to shed their leaves, the dry, east wind snatching them from the hedgerow and leaving it bare, and I found myself working beneath featureless grey skies as I pleached and bound the last of the blackthorn at Eweleaze. It had been a good start to the season. I had not lost any days through inclement weather and had managed to finish the work in good time. I was aware the prolonged dry spell had caused problems for the surrounding farmers, and many of them had been feeding their livestock precious winter feed for many weeks already, in the absence of sufficient grazing. The settled weather had however made my task here considerably more comfortable. The firm ground conditions had made it easier both to access and to work on the hedge and the pushing up and burning of the brushwood would barely mark the pasture. The difficulty of any outside labour is amplified greatly when the task is accompanied by driving rain but more especially, mud. The radio in the wagon was forecasting that the weather would be coming from the south-west in the coming days; this in turn would, at last, bring rain.

I was now spending the greater part of the evenings inside, lighting the stove most nights. Although the days were still pleasantly warm, there is a comfort to be had in a fire and with the retreating light I was grateful for it. I had stored a supply of timber under the wagon, gathering enough over the weeks to keep the stove going between contracts. Hedgerows have always provided a ready supply

OCTOBER

of firewood for their owners, and the larger lengths of ash had been collected by tractor for the farm's wood stores. The smaller, knottier lengths of thorn and maple were more difficult to process, and with the farm's wood stores well stacked, I was welcome to these less desirable lengths for my own fire.

All of the wood I had gathered was well seasoned and suitable for immediate burning. Blackthorn wood made up the bulk of the haul, though hawthorn, maple, ash and several pieces of oak lay amongst it. The oak had been a good find. It would burn particularly hot, and like coal it would keep the fire in the wagon ticking over throughout the night. I had found it nestled at the base of the hedgerow, the sap wood long since rotted away to leave only its bone-like heartwood, each piece resembling a stag's antler. The other woods, too, were all hot burning. I split them down into halves and quarters using an old ash-handled hatchet that was amongst my favourite possessions. I had purchased it some years earlier from a craftsman who lived in the Welsh borders and of whose work I was particularly fond.

Once split, the blackthorn's heartwood, like its bark when oiled, was an attractive wine-stain purple and it seemed almost a shame to burn it. If turned on a lathe it would make an attractive bowl or candlestick, so I decided to save a few of the prettiest pieces, in case I met someone who could make use of them in this way. I had already held back several lengths of thorn that I planned eventually to fashion into walking sticks, and hoped these would reveal the same colouration when I came to work on the handles. The mauve heart of the blackthorn can't be gleaned from the outside, and the discovery of a particularly attractive pattern and colouration is the hedgerow equivalent of a pearl.

I store only as much wood as I am able to travel with, which amounts to around ten days' worth when fully laden, perhaps slightly less. I rely on the stove not just for warmth but also for drying both wet clothes and wet wood. Wood collected during spells of wet weather will often be waterlogged and require several hours' resting next to a hot stove before it's suitable to burn. On these wet and windy nights the interior of the wagon has a distinctly busy feel. Items of clothing hang from hooks screwed into the oak beams on the ceiling, woollen hat and socks, corduroy trousers and shirt hang close to the stove pipe and steam. Saturated wood surrounds the stove and fills the wagon with a sweet autumnal aroma as it gently dries. Candles sit burning in old jam jars, and along with the oil lamp provide sufficient light with which to cook, read or write during the long evenings. And the teapot, of course, is always nestled on top of the hotplate.

It was time to leave Eweleaze. After packing down the wagon, I walked the length of the newly laid hedgerow,

OCTOBER

casting an eye over my work for a final time. I felt a sense of achievement and satisfaction in the completion of the contract; the unruly thorns had now been brought to heel and would once again fulfil their original purpose. The wind, as the radio had predicted, now came from the southwest and had lost its edge, although the clouds that accompanied it grew increasingly heavy.

I had my satchel with me, and before leaving I would take my share of the hedgerow's autumn bounty. The hedge had already provided me with a good supply of wood, but now I would harvest its fruits. Like the blackthorn's heartwood its fruit, the sloe, is a deep mauve colour, resembling a small plum. Sloes are bitter to the taste if eaten straight from the hedgerow, but despite this they are sought by birds and country people alike, and are a favourite of the mistle thrush in particular. Once the first frosts have sweetened their flesh and softened their skin sloes can be added, with a generous helping of sugar, to gin. Sloe gin is traditionally enjoyed around Christmas and the darkest days of midwinter, by which time the fruit will have infused the clear spirit and turned it a warming crimson. I pour mine into a demijohn and for the first week turn the vessel every day to ensure the flavours combine thoroughly. After that, I turn it at least once a week, and by the time we are on the cusp of Christmas, it is ready. This gift from the hedgerow is best enjoyed around an open fire on the clear, cold starlit nights at the turn of the year.

The previous night the landowner had joined me for tea and we had agreed to talk again over the summer to arrange the following season's work. On leaving she had presented me with a bottle of sloe gin and a dozen eggs from the hens in the orchard. I had given her a walking stick of hazel that she had admired on a previous visit.

The old diesel engine shook the whole cab of the wagon as it turned over reluctantly several times before coming to life. An alarm and flashing light on the dashboard reminded me I needed to allow the air pressure to build in the wagon's cylinders in order for the brakes to work; this takes several minutes and the lorry juddered rhythmically as it idled. Eventually, the exhaust smoke cleared from black to a grey blue, and I was on my way. The rain began to fall steadily as I pulled out of the farm entrance and on to the main road, the perishing rubber blades of the windscreen wipers smudging rather than clearing the rain from the glass. I had several days before I was due to start the next contract and decided to head south to an out-of-the-way pitch I knew of, south of Blandford, just a couple of miles away from the copse I had worked some twenty-five years previously as an apprentice. I stopped in Blandford en route for provisions; the cooler weather meant I was able to store fresh dairy and meat products, to some extent, so alongside the usual bread and tinned meat, I collected fresh milk, a modest block of cheese and some bacon, as well as several bottles of gin to add to the sloes I had collected that morning.

As I passed the gateway to the old coppice I couldn't resist pulling over to see the condition of the woodland. It had been many years since I had last seen old Bill Bugler, the woodsman who had first employed me as a young man, and I wondered if he was still working the woods, as he would be in his mid-seventies by now. As I climbed the rusting field gate into the coppice I was disappointed, although not surprised, by the scene before me. The hazel that made up perhaps a third of the woodland had not been cut since Bill and I had last worked it over two decades earlier, and it now grew tall, casting continuous shadow across the woodland floor.

Like hedgelaying, the cultivation and periodic harvesting of hazel is a regenerative practice. By cutting the tree at its base, it responds by sending up new growth from the stump, or stool, and so provides a sustainable supply of rods and poles whose length and diameter can be accurately predicted by the time left between harvests. Clearing in this manner also mimics the natural cycle of windblown trees, allowing light to the woodland floor, and by working separate sections of woodland annually the varying degrees of light and cover provide different habitats for wildlife and plants, increasing the woodland's biodiversity. Coppicing is an ancient technique that has been used by woodsmen for millennia, through the Bronze Age, Roman and Saxon periods, and continued to remain an important part of woodland management into the last century. As well as hazel, sweet chestnut, willow, hornbeam and oak have all traditionally been managed by coppicing. But as with the laying of hedgerows, coppicing has declined dramatically in recent times. And as the skills and knowledge used to work and preserve the ancient woods are being lost, so too is the biodiversity that relies on them.

I took the time whilst in the coppice to find and cut a couple of thumb sticks. I noticed each long, straight shank first before following it with my eye to a fork of even proportions which would make a suitable handle. I didn't spend long searching amongst the yellowing leaves of the hazel, however, as the canopy closing in above me felt oppressive in the late-October twilight. The rain had gathered some momentum by now and I pulled my collar up to meet the back of my cap. Amongst the trees, with the leaves falling steadily and the cold rain in my face, autumn felt amplified.

A thicket of bramble was growing alongside the gate back out on to the lane. By now most of the blackberries were gone – the earliest fruit had already been sweet and edible in August – but I still managed to pick a good pocketful before exiting the woodland.

OCTOBER

It took less than ten minutes to make the journey down the winding, narrow lanes to my pitch for the next couple of nights at Bulbarrow Hill. From here I could see Blackmore Vale laid out before me, with Somerset to the west and Wiltshire to the north. Among the hedgerows and woodlands, I could make out the vale's farms and hamlets. Many of the fields had been freshly ploughed and drilled, enhancing the patchwork nature of the landscape. A flock of seagulls busied themselves around a tractor, foraging for the worms and grubs cast up by the plough. A rook passed overhead, croaking. It tilted its head to the side to observe me before adjusting its flight when it realised it had caught my attention, gaining altitude quickly and sweeping away across the hillside.

It was now shortly after six in the evening and the light was beginning to fade. With only a couple of nights here I had decided not to unpack my belongings from the wooden trunk and had just the essentials to hand. I left the oil lamp stowed away and the paraffin it needed locked in the wooden tool chest under the wagon, and by the light of a couple of candles and the flame from the stove I prepared and ate a simple supper of hot soup and buttery toast. The wind was picking up outside but I left the door open. Occasionally a vehicle passed on the road outside and bright headlights momentarily illuminated the interior of the old wagon through the open door, highlighting the rain outside which now drummed constantly on the tin roof.

Gradually, the rhythmic patter coupled with the soft candlelight relaxed me. The traffic on the lane subsided, and I was alone on the hill. Through the window above my desk I was able to make out the main routes across the vale from the distant glow of the orange streetlights, while isolated pockets of bright silver light shone like land-bound stars, marking the illuminated farmyards. In between the lights the patchwork of fields was indistinguishable, and the copses and hedgerows were now lost to the darkness.

As the evening went on, I sipped my tea in the old armchair and reminisced on the time I had spent with Bill Bugler. Before meeting Bill my main employment had been as a seasonal farm worker and I had little experience of woodland management beyond some casual work cutting firewood. As a young man I had travelled between jobs, largely suiting myself and relying on no single trade to earn my living. However, my life had changed in the months leading up to meeting Bill. I had recently become a father and had moved with my daughter's mother into a cottage in a West Dorset village. I felt keenly aware of my new responsibility and although I had obtained no qualifications from school, I wasn't shy of hard work and felt confident I could find regular employment.

Initially, I had sold firewood and taken on odd jobs in the village, and although this had enabled us to get by, there was little consistency in this kind of work. I had come across Bill's contact details in an old telephone directory which had been left by a previous tenant of the cottage and had called him one evening. A softly spoken Dorset native, Bill told me that he was a small-scale contractor, selling firewood and coppice produce from the woodland he worked. He ran an old tractor and a saw bench from a clearing in the wood and in his own words wasn't 'trying to break any records'. Beside the hazel coppicing, there was a stand of beech that needed thinning, and beyond this some Scots pine too; in between this work there were some lengths of hedge to lay as well, so he would be grateful for my help. And so, aged just twenty-one, I embarked on my journey into both woodmanship and fatherhood. Sitting in my armchair nearly three decades later and looking back, I reflect that I was really no more than a boy.

I recalled my first day, meeting Bill in the car park of the Shire Horse pub in a village a few miles to the north of here, before following his pickup along the lanes and through the gateway I had climbed earlier. The first job of the day was always the lighting of the fire. Bill cared little for affairs outside the district, but kept a close eye on the horses and was never without an old copy of the *Racing Post* to serve as kindling. After placing his smoke-blackened kettle amongst the flames, we would descend into the wood for an hour before stopping for tea. It was during these tea breaks, in a smoky woodland ride, drinking from chipped, tannin-stained mugs, that my education began.

Bill would talk of the woods and countryside and the

changes he had seen. He often lamented the demise of the hazel coppice and the men he had known who had worked it. Many of the trades that relied on the woodland had become expendable, and so after centuries of careful management, large tracts of woodland had become redundant and now only a fraction of these old countrymen were still working.

As my work with Bill progressed, I came to meet many of these men who still made their livelihood practising the traditional crafts and skills that had been passed down, often from their fathers. I always listened intently as this multi-generational knowledge was shared generously around that fire in the woodland. I met the thatcher who came to collect the 'gads' we had cut from the coppice. These were lengths of hazel that he would split into quarters with a billhook before sharpening each end. He would take these up on to the roof he was thatching, where he would twist them into wooden staples that he used to secure the reed. I met the hurdle maker who split the rods we supplied and wove them into panels. The sheep hurdle had at one time been a mainstay of the hazel coppice and huge numbers of them had been manufactured each year to be used as pens and races; although they were no longer required for this purpose, there was still a demand for them as garden fencing. Firewood merchants arrived to collect the lengths of beech we cut, which they would split into logs and sell in the surrounding villages. The charcoal burner, with his blackened face and overalls, would site his mobile kiln in a corner of the wood and convert the larger lengths of hazel into barbecue fuel which could be sold to the numerous summer visitors who frequented the Dorset beaches.

It was also Bill who taught me to lay my first hedge, a roadside boundary of thorns that billowed into the lane from the bank on which it grew. He laid it in the Dorset style, pleaching each stem before lowering them gently until they came to rest on the top of the bank. I worked behind him, tidying the structure that once laid rose around two feet from the ground. Bill had brought several bundles of hazel rods from the woodland and I had used these to fix the pleachers into position, pushing the thickest, sharpened end of each rod into the top of the bank before bending it across the width of the hedge and securing it on the other side by tucking it under a pleacher. The hazel rod, or bond as it is called, now held the newly laid stem firm and prevented it from springing out of place or coming away from the stump completely. As each bond was set in place, Bill instructed me to cut away or tuck in any branches that jutted out of place and ruined the neat lines of the hedge. He stopped often to inspect my work, pointing out any bonds I had failed to secure properly and which had allowed the pleachers to sit up. 'If a terrier can get underneath 'em, they is too high,' he said. Nearly thirty years later I still use this rule when laying hedges in the Dorset style. The completed structure was a continuous half-barrel secured every couple of paces by a hazel bond, a river of thorns it seemed almost to flow along the top of the bank. The Dorset style of hedging was my introduction to the craft and remains a style I work in every season. All these years later, I remember those early days with Bill and the patient guidance he offered as if it were yesterday.

As I prepared for bed, less than a mile from the old woodland but some thirty years removed, I considered my responsibility to the hedger's craft. I was now around the age Bill had been when he first introduced me to the skill. Though I had become used to the solitude of the hedger's work I was keenly aware of the long line of woodsmen before me, and it seemed important that I, too, passed on these traditions. To do otherwise would be to the detriment of the land. My responsibility, I concluded, was not only to labour amongst the thorns, but also to teach.

4

November

At the turn of the month I left the pitch on Bulbarrow Hill. The retreating evening light had been shortened further by the end of British Summer Time, marked by the turning back of the clocks. Although this meant the mornings lightened earlier, sunset would now fall just after five. I woke on my last morning on the hill keen to make my way northward, away from the vale and on to the next contract where I hoped to get my new pitch in order before I lost the light.

I had been woken by the unmistakable sound of the tractor-mounted hedge cutter, its heavy iron hammers each armed with a sharpened edge spiralling the length of a steel cylinder that rotated at terrific speed. I could hear the tractor approaching down the lane that led to my stopping place, its engine whining loudly as its power was diverted into the cutting machine. Looking out, I could see the woody stems of the thorns being thrashed mercilessly by the hammers, while the heavier and older growth in the hedge was catapulted fiercely into the steel guard that surrounded the rotating cylinder, its steel cowling dented and showing only the faded and chipped remains of its bright yellow paint.

It was a grey, cheerless dawn, a thick blanket of low cloud rolling unbroken across the November sky and accompanied by a cold autumnal drizzle that largely obscured the view across the vale. But despite the hedge cutter working in the lane, as I sat on the steps of the wagon drinking tea it was still possible to make out the sounds of the countryside below. The barking of a dog and the lowing of cattle from one of the farms, then a distant voice which echoed across the vale, a stockman calling to his animals, perhaps bringing livestock into winter housing now that the weather was beginning to turn. Church bells pealed out, calling the residents of the vale to the Sunday service, although many of the pews would likely remain empty. Rural communities once centred around the church but in recent decades congregations had dwindled and fragmented. The bells, however, still rang out each Sunday, providing an accompaniment this morning to the metallic chiming of the tractor flails cutting the roadside hedgerow.

With the autumn cultivation work now complete on many of the farms, a good number of agricultural contractors would be turning their attention to the maintenance of the hedgerows. The pleaching and laying of a hedge is generally only carried out every fifteen or twenty years, and in the years between the hedgerow needs to be trimmed to prevent it from growing too high and shading out any adjacent crops. Trimming encourages horizontal rather than vertical growth, bulking out the structure and improving its function as a barrier to livestock. As the hedgerow becomes denser with each season, wildlife use it as a refuge, impenetrable to larger predators who, like the cattle and sheep, are discouraged by the entanglement of thorns.

Traditionally, the preferred method was to trim the hedgerow with a pitched top, the thick, impenetrable wall of thorns at the base narrowing to a ridge at its peak, ensuring any winter snowfall could be easily shed and thus avoiding damage to the boundary. The dwindling rural labour force in the second half of the last century saw this practice decline, and in some instances saw trimming abandoned completely. Left unattended for decades, these once meticulously maintained boundaries have reverted to trees and it is only their unnaturally linear pattern that speaks subtly of their past. Amongst these abandoned rows the keener eye may observe that many of the trees have distorted trunks, unnatural horizontal sections of considerable girth that give rise to multiple stems which now tower skyward. To those who recognise these signs, this sculpted growth speaks of the men who worked here in decades past with axe and billhook, pleaching and laying the stems with hand-forged tools, and harks back to an age when man's relationship with the land was closer, more intimate, gentler.

In the absence of a reliable rural workforce, and with the increased mechanisation of agriculture, the tractor-mounted rotary flail was developed. Although often demonised, it fulfils a role that has always been integral in a hedgerow's management. Used sympathetically to trim appropriately sized material, the flail is an essential labour-saving tool that is perfectly suitable for maintaining thick, healthy boundaries; indeed, I have met many operators who take great pride in ensuring the hedge is maintained to an exceptionally high standard. However, unless pleaching and laying are also incorporated into the hedgerow's management, continuous flailing will gradually see the boundary decline and die back, and over time the hedgerow will become unfit for either the retention of livestock or as a wildlife habitat. Recent years have seen the maintenance of hedgerows falling ever lower down the rural agenda, and as a result many of the remaining hedgerows are maintained by trimming alone.

Encouraged by regular trimming the hedge will initially become thicker, but eventually the growth at the centre of the dense thicket will become starved of light and begin to stagnate. Gaps will start to appear at the hedgerow's base and these will become exaggerated still further by the attentions of livestock, who by their nature will routinely test the barrier. These gaps are a sign that the hedge needs to be allowed to grow tall for several growing seasons and then laid once again, essentially starting the life cycle of the hedgerow afresh. The last few decades have seen these signs largely ignored, with the majority of hedgerows being cropped annually until eventually the solid lengths become broken lines of scrub which are then patched up with barbed wire and galvanised stock netting. These remaining lengths of scrub, which often serve as a last bastion for the declining wildlife on many farms, continue to be trimmed until eventually it makes more sense to remove them altogether, grub them out and install wire fencing instead.

As a result of this, the rotary flail has become a symbol of unsympathetic farming practices, of the lost partnership with nature which was previously so prevalent. The truth of course is far more complex, and farmers should not be demonised for the circumstances that unfolded in the rapid modernisation of agriculture. There are plenty amongst their number who lost their battle to sustain a way of life which was, in most cases, all the family had known for generations.

It was shortly after nine when I pulled away from the pitch at Bulbarrow Hill and made my way north toward the Wiltshire town of Mere. It was a Sunday and the roads were quiet, so despite stopping for provisions in Shaftesbury it was not yet midday as I passed through Mere, a quiet

Wiltshire town which sits on the border with Dorset, just south of the A303. This ancient trunk road connects the rural south-west to the more urban communities of the south-east, passing through Devon, Dorset, Somerset and Wiltshire before leading eastwards into Hampshire.

My next contract was located just north of Mere and comprised several hundred metres of mixed hedgerow in a state of decline. Like Eweleaze, Windwhistle Farm had previously belonged to the Duchy of Cornwall and was now under new management. Although I had liaised with the owner regarding the contract it was the farm manager, a young lady named Hannah, who I was to deal with and who I met on arrival. I left the wagon in the yard and Hannah guided me into the farm office where her dog, which had been sleeping beneath an old armchair, sprang excitedly into life, unable to contain its joy at being reunited with its owner who had been out working with the cattle for most of the morning. The handsome springer spaniel ran excitedly about my feet, wagging its docked stump of a tail enthusiastically and demanding my attention.

Eventually the dog settled and allowed us to drink our tea and talk. Hannah explained that she now held the grazing rights on the farm and ran a modest head of beef cattle on the land. She and her husband, who ran a timber-framing business from a converted shed in the yard, had been on the farm for eighteen months overseeing the contract work that had been taking place. The new landowner, who was currently away on other business, wanted to increase biodiversity on the farm and alongside the regenerative grazing techniques that Hannah employed with her cattle to restore and improve the pasture, had plans afoot for significant tree planting and the restoration of the hedgerows and water courses.

After tea, Hannah showed me the location of my pitch for the duration of the contract. I was pleased to see it was on hard standing to the rear of the yard and would be sheltered from the worst of the weather by a Dutch barn that was stacked to the rafters with straw bales from the summer's harvest and would be a convenient buffer from the autumn gales. Next to the barn was an old cattle shed, still fitted with rusting old cow cubicles from the farm's past as a dairy, which would provide further shelter as well as a water supply. An old brass tap on the side of the building drew water straight from the mains and, I was informed, rarely froze. This was good to know, as the contract would lead me through November and into December beyond.

With everything in order regarding the contract, Hannah bid me good luck and left me to get the wagon in place and organised. She had cattle to attend to, and as the drizzle became more persistent I was keen to get the wagon in a suitable position and the fire going before the light faded. Although it was only a little after two in the afternoon the overcast sky made the day feel later than it was and I lit the oil lamp, which gave off a homely glow inside the wagon. By shortly after three, the living space was in order and the teapot was brewing on top of the woodstove. As the light faded, I stood quietly by the wagon door, watching a pied wagtail who had now come to inspect this new resident to what was apparently his corner of the farmyard.

The evening drew in quickly and the drizzle that had persisted throughout the day settled into a relentless rain. At intervals it became torrential, creating a tumultuous din as the rain violently lashed the corrugated-tin sheets of the Dutch barn and the wind snatched at a loose sheet on the building's roof, causing it to clatter loudly as it bucked against the gale, threatening to tear loose. Then, as suddenly as it started, the rain would ease back to a steady, less aggressive patter as the squall continued on its path eastwards across the downs, leaving behind the noise of a constant flow of water hitting the concrete of the yard floor as it flooded rapidly from the failing guttering of the barn. As that, too, slowed I stood at the door of the wagon with the upper section open to allow the evening air in. I could hear the peep of the wagtail from within the shadows of the cattle shed. We had both managed to find shelter from the worst of the storm.

Shortly after five the following morning I was out of bed and preparing tea. Outside, it was still pitch black and the wind whistled as it blew around the wagon and the outbuildings in the yard. If I hadn't had a way of telling the time, the hour could easily have been mistaken for one much earlier. I find it harder to rise at this time of the year; there is no early-morning birdsong, no glow in the east from the window, and the only sound that reached me as I descended from the sleeping compartment was the wind and the old tin sheet still clattering, although not as urgently as it had the previous night. The rain had largely abated, blown through on the brisk south-westerly that was now taking the last of the leaves from the hedgerows and the acorns from the oaks in the fields beyond the yard. It was not cold inside the wagon, though we were approaching the turn of the year; the stormy weather had swept in from warmer climes to the south and, although turbulent, had brought with it the mild temperatures and moisture it had gathered on its journey across the Atlantic. There was no need to light the stove, and as the time approached six I was comfortably seated and drinking tea by lamplight with the stable door ajar, waiting on the approaching daylight.

The hedge to be worked on was thankfully only a short walk from the wagon. Across the yard, where the hard ground met the soft, grassy bank of a shallow ditch, were the remains of an old stock fence, its steel wire hanging loosely between decaying stakes. Despite carrying a chainsaw, a satchel of tools and a full fuel can, I was easily able to step over the slack wire and access the hedgerow by way of a deer track that led across the ditch and through a gap in the hedge that I would be laying. The hedgerow tracked uphill in a southerly direction for perhaps three hundred metres, carrying on past the farmyard and heading north for a further one hundred. Once I had scrambled through the gap in the hedge, I made my way along its length, climbing the slope to the farm's southern boundary where the hedgerow began. I planned to start here and work my way downhill, working from left to right so that the pleached stems were correctly laid uphill.

Upon closer inspection I found the hedge to be typical of other agricultural hedgerows I had worked on in the district. The primary species was hawthorn, which does not sucker as blackthorn does, and so at least the hedge had not widened as much as the one at Eweleaze; this, thankfully, meant I would spend less time on the initial clearance and preparation of the row. The plants themselves, however, were in a similar condition. Years of flailing had seen hard knuckles of scarred wood form where the hedge had been cut at the same height annually, around six foot. These knotted crowns were so entangled with both themselves and their neighbouring plants it would be impossible to unravel them, so each one would have to be cut with a chainsaw. The growth immediately below had also become twisted and warped and evolved from a single, somewhat sickly-looking, stem.

Among the hawthorns grew the usual field maples, although in this case it would unfortunately not be possible to either lay or leave them to grow on as hedgerow trees. They, too, had become heavily distorted and formed a dense nest of gnarled stems that could only be restored by coppicing them back to the deformed, bulbous stump at the hedgerow's base. With the extra light afforded by the laying process the subsequent regrowth could then grow straight and would likely provide suitable stems for the hedger to work with when the time came to lay the hedgerow again.

There were other species present, too – elm, spindle, blackthorn, hazel and elder had found a place here – but all were being overwhelmed by the briars. Although brambles are welcome in an agricultural hedgerow and reward both wildlife and hedger every autumn with their fruit, once a hedge is in decline as this one was, they will rapidly take over. In time they will come to dominate the row, smothering the faltering thorns, and eventually leading to the hedgerow's demise.

The hedgerow here was to be laid in the Dorset style. As is common with many of these West Country boundaries, the bank on which the hedgerow had been originally planted had slumped. This had tempted livestock to climb the bank in order to browse the hedgerow plants, something sheep in particular are prone to. The problem had become exacerbated by water run-off and eventually the earth bank had subsided completely, and now provided no barrier at all. A complete restoration of a West Country hedge would see the bank renovated, or 'cast up' to use the regional term. The slumping soil from the bank would be re-faced with a sharp shovel, and the spoil placed on top of the bank covering the newly pleached stems, the freshly dug earth encouraging the plants to put down roots.

Here at Windwhistle Farm where there was still plentiful vegetation, I decided to pleach and lay the thorns despite the loss of the earth bank. This would reinvigorate and increase density in the hedgerow plants and see a healthy hedgerow re-established in the coming years. Future laying could then be carried out in one of the more upright styles such as the Midland or South of England, where the hedgerow is supported on stakes and bound with rods of hazel or willow. Although neither of these styles are traditionally carried out in the district, this would nevertheless see the re-establishment of a thick and healthy hedgerow which would be of huge benefit to the local wildlife as well as being capable of retaining livestock, and is preferable to either leaving the declining hedgerow unmanaged or removing it completely in favour of fencing.

As we work to restore as many of Britain's hedgerows as possible, laying only in the regional style may be preferable, but it is not always feasible. These boundaries were originally created according to the materials to hand and the workforce available, and this has changed over the centuries. In my opinion, and although perhaps upsetting to some traditionalists, at a time when we have seen such a drastic decline in farmland biodiversity directly linked to the loss of many miles of hedgerows, we must use what skills, tools and time

NOVEMBER

we have to reinstate these wooded boundaries as best as we are able. Although preserving regional methods where possible, we must not be afraid to introduce new ways of thinking; in this way traditions are built upon, evolve and survive. A traditionally managed West Country bank and hedge is a structure of great beauty and there are, thankfully, still many in a satisfactory condition which continue to maintain the individuality and heritage of the district.

By one o'clock in the afternoon I had made a good start on the southern boundary at the top of the hill. This length of hedgerow contained several large open-grown oaks, their spreading and well-proportioned crowns now in full autumnal colour, having turned later than the skeletal thorns below. The clouds rushed across the sky in the wake of the previous evening's storm and a break in the cover allowed the November sun to briefly catch the oaks' canopy, emblazoning the golden leaves.

In the shelter of one of these old oaks I lit a small fire, preparing a bed of dry hawthorn sticks over some newspaper before placing a couple of old candle stubs and some dried orange peel amongst the kindling. The wind made it difficult to keep the match alight long enough for the flame to catch, but eventually I succeeded and the breeze now helped to blow the flames rapidly through the kindle wood, which crackled loudly as it began to combust. The previous evening's rain had left the lying wood saturated, but I had found plenty of standing deadwood amongst the thorns during the morning's cutting, which not only added fuel to the fire but also acted as a combustible trivet on which to balance the old kettle.

Despite the wind, the ground was still damp and I sat on my old wax waistcoat to relax for half an hour in front of the fire whilst eating my lunch, enjoying the smell of the woodsmoke and the early-afternoon sunshine. The hawthorn burned hot and it didn't take long for the old kettle to start whistling, a most welcome noise when out working in the fields during the autumn. I poured the scalding water into my stained enamel mug, the inside darkened from the tannins of many a cup of strong and smoky tea enjoyed outside next to the hedgerow.

I was not alone under the oak. Throughout the morning I had been accompanied by a wren and a robin. The brazen robin worked within feet of me and despite the din of the chainsaw had been confident enough to pick through the leaf litter that I'd disturbed while clearing the base of the hedgerow. The wren was a little more cautious and preferred the brush pile that had been accumulating opposite the hedge. He had worked his way along the discarded stems and through the ivy and bramble which were bound for the bonfire, picking out the odd morsel. Taking advantage of the lull in proceedings over lunchtime, he was now working the leaf litter of the hedgerow, skipping along the newly pleached stems and occasionally darting beneath them before nervously retreating once more to the brush pile.

The robin eyed me curiously and skipped along the ground just yards from where I was seated. The robin is a constant companion throughout the winter months and I found myself wondering where he had sheltered in the storm last night, if he had seen winter before and if he was strong enough to survive the season to come. As this thought crossed my mind I threw him a crust of bread primed with a little peanut butter, but only succeeded in startling him and he took flight in alarm.

As the month progressed, the ever-retreating light was subdued still further by leaden skies and often a dense fog which lingered on the days when the westerly wind abated. On a still morning late in the month the fog had engulfed the fields and hedgerows, stifling the sounds of the country around me and bringing with it a feeling of isolation, of being cut adrift from the modern world. Just a few miles away a constant flow of traffic rushed along the A303 between the towns and cities of the south and on clear days the melee of freight and commerce was clearly audible from the farm, but today the fog silenced everything, narrowing the world to where I worked amongst the hawthorns, with only the occasional rasp of a lone rook passing overhead or high-pitched shrill of the wren or robin for company. It had been some days since I'd had any interaction with another person and alone with my thoughts in the murk, a strong sense of loneliness came over me.

In an attempt to lift my mood I reminded myself of the value of my efforts for the landscape around me, but as I cast an eye over the morning's work, I found little satisfaction in what I had accomplished. Despite a continuous row of thorns to work with, the pleached stems had laid awkwardly and seemed to sit untidily between the oaks, which by now had lost their autumnal vibrancy and appeared grey in the November half-light. A hedger can only work with the stems available and on a restoration such as this, the initial cutting is less about the aesthetics of the build and primarily about reinvigorating the plants that remain. The hedger must look several years ahead and envisage the hedgerow restored, once again sturdy and with plentiful growth. Today, that eluded me. My hands were stiff and sore from the sharp thorns that had passed easily through my tattered gloves, and despite making good progress with the contract I was finding it difficult to maintain enthusiasm for the work.

As a steady drizzle began to fall, I decided to return to the wagon. I had reached the stake marking that day's objective and was lacking the drive to press ahead any further. I take the days as they come and although I always

NOVEMBER

ensure I gain some ground I see little benefit in staying out in poor weather on days like today. I was making steady progress and would complete the contract in good time. On returning to the wagon, I found the familiar wagtail skipping merrily about the yard. He paid no mind to me and carried on with his tasks, and watching him I was reminded of the words of D. H. Lawrence, 'I never saw a wild thing sorry for itself. A small bird will drop frozen dead from a bough without ever having felt sorry for itself.'

Once inside the wagon I filled the kettle from the water that remained in the enamel jug and put it on the hob, then got a good fire going, leaving the door on the wood burner open just a crack to allow the hungry flames to draw in the air and build rapidly. I hung my jacket and hat behind the stovepipe to dry and by the time the kettle had boiled the room was already warming up. A small battery-powered radio provided me with some company as I warmed myself through by the stove, and I was able to enjoy my tea in asome comfort.

By a little after five the light had all but gone but thankfully my mood had lifted to some extent as I prepared a thick stew with some tinned steak. I don't always eat as well as I should, and when tired and hungry often opt for whatever is quickest: tinned soup or meat, bread or some eggs if I have any. I am aware, however, that such physical work, outside in all weathers, requires a certain amount of nutritional sustenance. Someone once told me that physical work outdoors keeps you fit for a long time, and then ages you very quickly. I have always relied on my body for my income, earning a living from working the woods and hedgerows. Now in my late forties, I am aware that my strength is finite and that I should do what I can to preserve it, as I hope to continue my work for many years to come.

The contract at Windwhistle was all but finished as November drew to a close. By now I had descended the southerly slope past the yard and was working northwards on the final hundred metres that would complete this section. The day was clear, although the sky remained a light shade of grey. It was dry, however, and with the end of the contract in sight I felt inspired to get the most out of the short days. As I cast an eye back along the length I had completed, a continuous drift of brushwood ran parallel to the hedgerow and marked my progress. This row had been in a particularly poor condition and so alongside the pleaching and laying which made up the majority of the restoration, I had been forced to coppice a significant amount of the plants.

Coppicing is perhaps the oldest form of hedgerow or woodland management; the hedgerow after all is essentially a linear woodland, a strip of trees and shrubs that originated in the understorey of the wildwood, capable of thriving in the dappled shade under the canopy. This understorey was exploited for wood and fruits by our hunter-gatherer ancestors and coppicing woodland became a method of ensuring a constantly regenerating supply. As our ancestors settled and hedgerows became established as boundaries, this ancient form of management was applied to them too, and although the boundary was unsuitable for use as a barrier in the years immediately following coppicing, by rotating grazing livestock between pastures this caused little inconvenience. In a modern setting coppicing a hedgerow still serves the same purpose. While laying can create a secure fence from the outset, coppicing still forms an important part of hedgerow maintenance and a hedgerow that has been left to become derelict can be reinvigorated by coppicing.

In the case of this contract, it had been used in conjunction with pleaching and laying to maximise the amount of regrowth for when the hedge next came to be laid; it is only at this point that a robust, stock-proof barrier will be established. My work here was just the first stage in a restoration that will likely take another ten years or so. At a time when the pace of living has quickened and immediate results are often demanded, this time frame may appear too long to some. But the countryside in which we live and work is a legacy that has been carefully crafted and passed to us by many generations before; our responsibility lies not only in what serves us today but what we pass on to the custodians of the land to come.

5

December

I completed the contract at Windwhistle Farm as the first week of December drew to a close. It had taken just shy of five weeks. Now I was making preparations to move on to the next block of work, heading west across the Somerset border. I had several smaller hedging jobs lined up, that amounted to little more than eighty metres in total, all within a short distance of each other. I intended to complete these jobs in the weeks leading up to Christmas before moving on to a large hedge-planting job in Dorset in the New Year. I had made prior arrangement to stay with friends at Marsh Farm for the duration of this work, a holding comprising some four hundred acres and lying on the fringes of the Quantock Hills. My old pickup truck had spent the autumn there, and I would use it to commute to the various contracts whilst enjoying the festive period with friends and family who live in the area.

After carefully navigating the narrow lanes that led to the

main road, I made my way westwards. The Wiltshire farmland peeled away from the road to both north and south and the vast, newly cultivated fields which had been sown with winter barley were a pleasing shade of green as the crop now rose from the tillage. The broad open acres were peppered with small woodlands, vestiges of the old wildwoods which were retained when the land was put to the plough, primarily for the cover they provide to pheasant and partridge. The land here was now more or less devoid of hedgerows, however. These had long since been grubbed out to accommodate the ever-larger farm machinery that struggles within our smaller enclosed meadows and is more suited to the vast plains of North America.

The old wagon slowed upon encountering the first of the hills I would have to cross to reach the flat lands of Somerset, and the drone from the engine beneath me increased significantly in volume. I took a brief glance in the rear-view mirror at the growing queue of traffic behind me before looking back out across the farmland around me. I found myself imagining how the landscape would have looked when the Wiltshire-born nature writer Richard Jefferies had walked these fields well over a century earlier, when the now-absent hedgerows would still have provided cover and sustenance for overwintering birds and linked the isolated woodlands across fallow fields.

As I progressed westward the landscape eventually became flatter and the road widened to two lanes, at which point

DECEMBER

I was overtaken briskly by a steady flow of traffic. At Yeovil, a town which lies close to Somerset's southern border with Dorset, I turned north and left the main trunk route behind me, delving into the narrow Somerset lanes enclosed within trimmed elm hedgerows. The landscape here felt old, timeless, and I had a great affection for it. Ramshackle farms and crumbling stone barns were easily visible over the hedgerows from my elevated position in the cab. Here and there were the ancient cider orchards characteristic of this region, their crooked apple trees wizened and bent by weather and age after countless seasons sustaining many a farm labourer with their annual crop. Among their branches hung clouds of mistletoe; it prefers apple trees above all others, although it will settle for oak, lime or poplar. Like hedgerows, these small cider orchards have an important role to play in increasing biodiversity, and are also a part of the district's heritage and a direct link to an agricultural past almost forgotten.

As I continued north to the Somerset Levels, the land changed once again. The earth here was a rich, soft peat and the road had sunk and moved, becoming uneven and causing the wagon to pitch and roll. The hedgerows either side of the road now gave way to deep ditches. Farmers have always used whatever lies to hand to contain their livestock and just as hedgerows were formed on the British lowlands and stone walls on the uplands, the easily workable dark peat of the Somerset Levels has been trenched and ditched to provide the barriers needed. The area is mazed with a network of ditches and drains that not only create barriers to livestock but also hold and disperse water throughout the year, keeping the land drained and providing summer grazing for livestock on its lush, species-rich pastures. The mosaic of habitats here supports a myriad of wetland birds and wildfowl, not least the heron's shy and rare relative the bittern, who finds cover amongst the reeds and sedges of the waterways. I have unfortunately never seen one, although I have a fondness for the grey heron who is found here in large numbers. It is from these wetlands that the county gets its name, Somerset being 'the land of the summer people', in reference to the early inhabitants who eked a living from the wetlands during the summer months.

The road was lined at intervals by ancient willows which thrive in the wet conditions here. These trees were once pollarded – that is, cut on a rotation not dissimilar to coppicing, but where only the crown of the tree is cut back. Pollarding willows produces a harvest of rods that would once have been used for wattle fencing, sheep hurdles and basket making, although these practices, once commonplace, have become became increasingly rare as cheaper plastic and metal products have replaced the traditional techniques.

The landscape grew increasingly wet and marshy as I neared my destination. While the water of the Levels drains into the Rivers Axe and Brue to the north, I was headed south to the marshy estuary where the tidal River Parrett

DECEMBER

disperses it into the Bristol Channel. My next pitch was at Marsh Farm, nestled between the Quantock Hills and the Parrett's muddy banks on the fringes of the village of Cannington, lying on the boundary of the district known as Sedgemoor.

Marsh Farm is owned by the Popham family, who have farmed in this part of Somerset for generations. Although it had once been a small dairy, like many others in the district it had recently abandoned milk production and the farm now fattened bull calves and kept a small herd of suckler cows instead, alongside a good head of sheep which grazed its pastures. A substantial amount of the farm was grassland from which a cut of hay was taken every summer as well as some silage in recent years, and a few fields were set aside for growing barley and corn. The farm now also ran several tipper lorries supplying local builders with sand and stone from a quarry that lay just beyond the village, and it was this business that kept the farm afloat. It is near-on impossible to run a small mixed farm and stay in the black; farmers are faced with having to diversify or be engulfed by larger agricultural business, usually sizeable arable operations or extensive dairy concerns. Amongst this changing rural environment, Tim and Susanne Popham have had to work hard, year round, to keep the farm viable and moving forward. I have a great deal of respect for the family, dogged country people, hard-working and fiercely protective of their heritage, yet warm-hearted and generous of nature.

Upon pulling into the yard I received a warm welcome and after exchanging pleasantries I was directed on to a patch of hard standing to the rear of the farmyard, where I had stayed several times before. The plot was surrounded by a ditch and enclosed by a hawthorn and elm hedgerow on the far side, with a stone track leading out to the pasture beyond. I parked the wagon in the shade of a large poplar tree, making sure to keep the rear of the wagon facing north and out of the prevailing weather.

The village of Cannington was just a short walk from here, no more than half a mile, and as the evening approached, I decided to make my way to an old inn I had frequented during my last stay. I had been alone a lot over the last few months, and was now keen to catch up with old friends, hear the news and gossip of village life, perhaps spin a yarn or two myself. Such is the way of country people.

It was dark as I walked through the farmyard and down to the village, and I used an old torch to light my way down the stone track, which was lined with a thick elm hedgerow and unlit. The cattle in the farmyard were undisturbed by my presence and carried on chewing hay in the shelter of the old barn as I passed. After a couple of hundred yards the loose chippings underfoot turned into a made-up road and the first houses of the village began to appear. I became aware of the smell of wood and coal burning on fires within the old cottages, and saw the smoke rising from the bricked chimneys into the clear, cold night. By now many of the cottages were decorated for Christmas, lights draped over low-slung thatch or around wooden porches strewn with muddy wellington boots. Outside one house the bare limbs of an ancient apple tree were hung with lights like icicles that gradually faded from white to blue and back again. Heavy wooden doors with pitted iron hinges were adorned with wreaths of fir and holly. On the way in I had noticed that the neighbouring farm was selling Christmas trees and wreaths from a temporary shed erected in the yard and seemed to be doing a roaring trade. Christmas was less than three weeks away.

As I walked on, I arrived at the village green, which was encircled with lights of all colours, bright greens, blues and reds trailing loosely between the cast-iron Victorian lamps. The inn lay beyond the village green and was nestled snugly into a row of cottages along Church Lane; the village church itself was just a hundred yards or so further along. I remembered to duck as I entered through the low oak door that had caught me out more than once, particularly on leaving. I was pleased to see a good fire of ash wood burning in the inglenook fireplace on the back wall of the public bar.

An ancient elm beam decorated with horse brasses supported the breast of the fireplace which was almost the size of the bus stop opposite the inn. The whitewashed stone chimney breast was blackened from centuries of woodsmoke that had escaped the cavernous fireplace and the room, low ceilinged and dimly lit, was redolent with the scent of smouldering ash wood.

I placed my jacket over a vacant bar stool before dropping my flat cap on the varnished wooden bar and placing the keys to the wagon and my wallet safely in it. Pulling out a twenty-pound note I caught the barmaid's eye and ordered a pint of Taunton cider, a dark sweet liquor fermented in old brandy barrels and held in high esteem within this area, although virtually unknown beyond it. A wizened old man was seated at the bar, his cap balanced precariously on one side of his head and a tangle of white whiskers escaping from underneath it and tracking down the sides of his ruddy cheeks. He smiled approvingly as I took a first, long draught of cider and we soon struck up a conversation. He introduced himself as Reuben, and it became apparent almost immediately that he had been in the pub for a fair few hours, judging by the drawl to his strong, West Country accent. As is often the case in these rural pubs, a stranger is initially judged on the drink he orders, and then, after some light interrogation, on his occupation. In these parts, being a hedger is deemed worthy of a seat at the bar.

In recent times, many rural pubs have become primarily

eating establishments catering almost exclusively for the more middle-class, wealthier customer. The local working man who spends his money simply on beer or cider, which offers scant reward to the publican, has to some degree been pushed out. Adding to this, the increased tax on beer has seen the cost of a pint, a staple for the working Englishman, become almost a luxury. However, in quiet corners of the countryside, tucked away on the edges of isolated hamlets and villages, traditional country pubs frequented by working countrymen can still be found. There, over a locally sourced brew of not inconsiderable strength, an evening can still be enjoyed by the true characters of the countryside. The inn on Church Lane was one such pub.

Once old Reuben and I had exchanged some brief words on the merits of Taunton cider he enquired, 'Ave er bin busy then or no?' I told him I had only just finished a good stretch of hedging yesterday and would be tying up loose ends here over the next few weeks before making my way to Dorset and a substantial hedge-planting job in the New Year. Reuben, it transpired, had spent a number of years working the woods himself and had laid many hedges in his time. We talked of work for a while before he told me a little of his life in the district. He had been born in an old cottage not far from where we were drinking, on the edge of a large wood that sprawled across the hills south of the village. He was from a large family, the youngest of seven brothers and sisters. He then attempted to name them, in numerical order: 'There be Joan, one. Maurice, two. Albert, dree. Mary, vour . . . No, no, Mary, dree. Albert, vour . . .' With some difficulty he managed to recite his siblings to me, explaining with some sadness that it was now only him and an older brother, Samuel, who remained.

I was enjoying listening to old Reuben and ordered two more drinks. The barmaid smiled as she placed our pints on the bar, and I suspected she had heard these tales on numerous occasions, though she obviously had great affection for the old man. Reuben had always been close to his brother Samuel, it seemed. He told me how as youngsters, and with living space at a premium, the two of them had shared a bed, though without enough bedcovers to go around they had made do with thick canvas feed sacks from the farm where his father worked. Despite this, he told me he had no memory of ever being cold or uncomfortable. As he took a long draught of his cider his eye landed on the holly tucked behind a distressed mirror which hung behind the bar, now casting little reflection though the old advert it bore for Scotch whisky was still visible. He began to reminisce on his childhood Christmases with Samuel. 'We'me would always ask for lead soldiers, uz boys,' he informed me. 'Uz always dry da get zum lead ivv uz could.' He went on to explain that the boys had little interest in playing with soldiers though, and that shortly after receiving the gift they would descend into the wood where they would make a small fire, melt the lead down and fashion the soldiers into shot, which was the best ammunition for the catapults they whittled from holly, hazel or hawthorn. 'Zamuel 'ad a reet vansee one ee vashioned vrom deer antler ee vound at the back of the wood.'

By now our glasses were empty once more and it was Reuben who ordered the next round, rummaging through the deep pockets of his well-worn tweed jacket before extracting a tattered five-pound note that looked like it had been there for some time, and 'da ztill bare the queen's 'ed'. We greeted some new arrivals briefly before he went on, telling me of the small game he and his brother were able to take with their slingshots: rabbit, pheasant, partridge, squirrel, even a hare or two, all for the table. Such skills which might provide sport for the more gentrified country dweller were more often born out of necessity for the rural working class, and have sustained many country families during lean times. 'Ave er ever eaten coot, Paul?' enquired Reuben, now much enthused as he reminisced about his younger days. A coot is a common water bird that frequents pools of still water, slow-moving rivers and streams; I often come across them in the large, water-filled ditches of the West Country. They have a habit of laying low in the cover by the water's edge and then bursting out when a person approaches, and making a swift dash for the far bank in a flurry of wings. The frantic flapping rarely sees them take flight, at which they are somewhat inept, but rather propels them in their sprint across the water's surface. They have often caught me off guard when I have unintentionally flushed them from cover beneath my feet. As of this time, however, I have never eaten coot. Reuben, it seemed, delighted in the culinary qualities of the bird and began elaborating on the various ways in which it could be prepared. 'Boiled, fried . . . Ifen ee get a chance, Paul, ee shold take un! Tis a lovely bit a meat on the bird.'

I passed the whole evening in conversation with old Reuben. Though I very much enjoyed listening to his stories I had by now consumed upwards of four pints of Taunton cider and being unaccustomed to such excess these days, and keen to warm the wagon through, I bid farewell to old Reuben and the barmaid. Saying my goodbyes I pulled on my jacket and cap and went to leave, thumping my head soundly on the low oak beam of the door. 'Go careful, matey!' Reuben exclaimed. 'Thik old door still catch I out too!'

The air was still and cold as I made the short journey back to the wagon. I had no need for the torch this time, as the clear night sky provided enough light despite only a thin crescent moon hanging above me. The cattle in the barn had settled as I passed through the yard and apart from the occasional deep lowing from the back of the shed and the rustling of straw in the shadows, it was quiet on the farm.

DECEMBER

As I neared the wagon, the call of a little owl echoed out from the crown of the poplar tree it was pitched under. Several seconds later, a reply came from the sycamores surrounding the old water mill which lay a couple of fields further along the track. I'd seen these stout little birds often whilst staying at Marsh Farm. They were confident creatures, accustomed to the activities of the farm. During the spring when the chicks were in the nest it was not unusual to see one of a pair perched upon an old fence post or peering down from a telegraph pole in broad daylight. I had often passed within yards of these birds and they would not take off in fright as you might expect, but would simply watch me pass before turning their gaze once more to the ground in search of the invertebrates or small mammals on which they fed.

The wagon was chilled through as I entered and I lit a couple of candles to illuminate the living space whilst getting ready for bed. The cold air and walk home had cleared my head and I felt hungry. I realised I had not eaten anything since arriving on the farm at lunchtime, and the time now was long past ten. It was too late to light the stove and prepare tea so I made do with some bread and butter and a mug of milk to settle my stomach, before undressing briskly and climbing into the chilled sheets, pulling the duvet up tightly around my ears whilst I got warm. After the long day and the evening's drink I was soon drifting off, snug in the sleeping compartment of my wagon, accompanied by the shrill cry of the little owl calling from the poplar above my head.

I was awake before first light on the Sunday morning. Although I was warm and comfortable in bed, I had woken with a considerable thirst after the previous evening's drinking with old Reuben which forced me to rise and put the kettle on to boil. The stars were still clearly visible against the blue black of the early morning but the moon had already left the sky which was lightening gradually away to the east. I dressed quickly in the chill and pulled on my woollen hat, then opened the door to see that a hard frost had descended overnight. By the time the kettle whistled, I had lit a good fire in the stove, and as the wagon warmed and the light outside increased, I set about the morning's tasks. I filled the enamel water jugs from the jerrycans in the cab, slivers of ice plopping into the jug. It was already warm enough to comfortably strip to my waist and wash in the steaming water of the Belfast sink before refilling the kettle for more tea. I prepared toast under the grill of the gas cooker and enjoyed it, heavily buttered, whilst leaning at the open stable door of the wagon. The cold air on my face felt refreshing after the hot flannel, and the heat of the woodstove behind kept me warm all the while.

After breakfast and with the fire now burning steadily

DECEMBER

and requiring no further attention, I stepped outside to enjoy the early morning. The sun had just risen above the pasture to the east and cast its rays, which in this late month gave only light and no heat, across the landscape before me. The hedgerows were devoid of leaves except those of the ivy and ever-present briars that rambled along their length, and the bare thorns glistened as the rays of light reflected from the hoar frost which clung to every part of them. The trees, oak, ash and alder, were transformed, every branch coated in frost. The sunlight split as it passed through the crown of the poplar overhead and the fractured beams caught the woodsmoke which rose almost directly upwards from the wagon's chimney in the sharp morning air. At intervals it was possible to make out the cattle in the yard lowing and I could hear the sound of the tractor in the yard firing into life as the morning's feed got under way.

As the month progressed, I was able to complete the various small hedging jobs I'd lined up in the surrounding villages. These jobs tended to be for customers living in old cottages with sizeable gardens that had once shared a boundary with an adjacent farm, and so were planted with the familiar thorny species used in agriculture but rarely seen in ornamental garden hedges. As awareness of the wildlife value of our traditional hedgerows has increased in recent years, these new cottage dwellers are more inclined to restore a traditional boundary rather than removing it in favour of garden fencing or ornamental conifer hedging, which has only a fraction of the environmental benefits of a mixed-species, deciduous British hedgerow. And so I acquire several of these jobs every season, and always try to fit them in around the Christmas period. At this time of year I am usually kept well-oiled with a regular supply of tea whilst I work the cottage gardens, as well as an array of seasonal treats: mince pies, some mulled wine, or a slice of Christmas cake of which I am particularly fond. In the spirit of the season I return the kindness. By the latter part of December, the sloe gin that has been curing in the wagon is good to drink and I pour a generous measure from the demijohns into clear glass bottles as gifts for each customer.

On these smaller cottage jobs, alongside the refurbishment of any older boundaries I often find myself dealing with lengths of newly established hedge. These are increasingly being planted in kitchen gardens to match the original boundary hedgerow and to act as a windbreak to the flowers and vegetables grown in raised beds. I prefer to carry out the work on these younger, more agreeable lengths using the traditional hand tools of billhook and axe. Generally, there is much less clearing out to be done and the vast majority of the vegetation can be worked into the finished structure. The bases of each stem can be easily accessed and are clean and supple; one well-aimed blow with a sharp

DECEMBER

billhook from the right hand whilst applying a little pressure to the plant with the left will see them pleach easily and lay tidily into position. Any unruly branches that ruin the line of the emerging structure can be pleached and folded into the construct, and by ensuring they sit at the same angle as the main stem, the consistency of the line is maintained and the hedger's craft accentuated. Once completed, with straight hazel stakes placed at regular intervals and bound tightly with the hazel rods, the structure becomes almost sculptural.

These Christmas jobs are a pleasure and the customers are never displeased with the completed work, which will often draw complimentary remarks from passers-by. Like many who work alone I am fond of conversation when the opportunity presents itself, and at this time of the year it is usual for people to stop and enquire about progress. I find it a pleasure to talk them through the work, explaining the technique and the thinking behind the method. It is encouraging to see their interest and I'm rarely short of words when talking of the benefits of hedgerows and the old skills required for their preservation. When the New Year rolls round I will have just the robin and the wren for company once more, and so for a few short weeks in December, I make the most of my time in the village.

Christmas Day approached, and my contracts for the month were completed with a few days to spare, so I busied myself on the farm. Having been working on only short lengths of hedgerow that month I had been unable to gather sufficient firewood to sustain the woodstove, but Tim had said I was welcome to help myself to several stems of dead alder that, although still standing, were already seasoned and suitable for burning. These alder stems lay beyond home field on the edge of the creek that crossed his land and on the morning before Christmas Eve, with the ground frozen solid, I was able to drive my old pickup truck easily across the fields to the edge of the watercourse.

The weather had remained calm and settled throughout December, and although bitterly cold, I found this much easier to cope with than the mild and wet conditions which see constant mud and damp encroach into all parts of my life. The mud, despite my best efforts, will always find its way into the wagon. The cooker and dresser door become scuffed with it as I enter with wet, muddy boots, and it gathers on the unit and armchair as damp clothes are sorted and hung to dry above the stove. It becomes a constant battle to keep the living space clean, and when spending many a long, dark week amongst the mud and thorns at the back of an isolated farm field, returning to a welcoming, tidy and well-maintained space is more than a pleasure, it is essential.

Today, however, was a pleasant morning, bright with cold

DECEMBER

winter sunshine. Rather than make several trips back to my pitch with lengths of timber I decided to process the firewood in the field next to the pickup truck, first cutting the alder rings before then splitting them with a maul and throwing them into the bed of the pickup. Despite the freezing temperatures the old adage 'firewood warms you twice' was true to its word and I was soon down to my shirt, perspiring as I swung the maul for a good couple of hours, the radio in the pickup playing Christmas carols between news bulletins as I worked. I left a couple of standing dead stems for the woodpecker, as old Bill Bugler had taught me when I had first started in the woods many years before. 'The old pecker bides 'ere just like we, always make sure to leave he a place for his hole,' he would say.

Eventually I had split enough wood to fill the pickup and the load rose pleasingly above the edge of the bed and against the rear window. I was pleased with the harvest; the dry alder would not only see me through Christmas but on to the planting job in January, where wood might be hard to come by. I chucked a couple of half-rotten lengths on top of the load and brought these back to the pitch too, as although unsuitable for the stove, if mixed with some dry hawthorn from the hedgerow these would burn well on the open fire outside the wagon.

I rose late on Christmas Eve. It was already light when I woke and noticeably less cold. The high pressure front which had gifted us the frost and bright sunshine was leaving, heading away to the north according to the weather forecast on the radio, and the weather would once again be coming in from the south-west, although the rain wouldn't arrive until after Christmas Day. There was no frost on the ground when I went to collect firewood from the loaded pickup, and the blue sky of the previous few days had given way to a light shade of grey. I was in good spirits; tomorrow I would be spending the day with my sister and her family, and it had been more than a few months since I had enjoyed her cooking. Her roast dinner, in my opinion, is unrivalled. My nieces and nephews would be there, as well as my mother and father who lived nearby. Family is important to me, and despite many months spent alone I always check in with them regularly. I have few close friends; I move around so much throughout the year I rarely keep in touch with the people I meet, although I do enjoy the company of old companions when a contract or summer pitch brings our paths together for a time. It is only family – my two beautiful daughters, my sister and her children, and my parents – with whom I speak regularly.

The following days were spent in much cheer, full of laughter and drinking, as is often the way when family reunite after many weeks apart. My sister's children are now grown up; the two boys have left home and work

DECEMBER

long hours in agriculture. The elder is employed on a busy dairy in Dorset, and had made his way home with his young wife and two daughters, having recently become a father at twenty-one, the same age as I was when my first daughter was born. My younger nephew is kept busy contracting near Bristol, but had also managed to get Christmas off. My two nieces were also present and it was a treat to see them, something I regretfully only manage infrequently these days. My own two daughters, Rosie and Lily, were spending Christmas with their mother in Dorset, although I did talk to them at length on Christmas morning and with my next contract taking me to that county we had arranged to toast the New Year together at the first opportunity.

6

January

The year turned. I had been pitched under the old poplar at Marsh Farm for almost a month and though the year was barely a few days old my thoughts had already turned to the planting contract in Dorset. The hedger's work encompasses several forms of management and alongside laying and coppicing, the planting of new hedgerows is an essential part of restoring biodiversity to our farmland. As a young apprentice years before I had been told that if a new hedgerow is planted late in the winter you have to ask it to grow, but plant it in the cold months of December or January and you can *tell* it. A hedgerow planted late in the season will suffer if followed by a dry spring but by planting earlier, in the depths of winter, the young roots have time to settle, giving the juvenile plants the best chance of success. When the weather is at its worst, when the rain drives in horizontally from the west and the ground is sodden and slippery underfoot, this is the best time to plant a new hedgerow.

And so, shortly into the New Year I left Marsh Farm. I had enjoyed my stay there; in some ways it was the closest I felt to home, under the poplar in the yard. Before I left, Tim agreed that come the spring I could return and use the pitch as a base to run my summer fencing work. Staying in one place for those months makes it much easier to store the large quantity of hazel rods needed for the wattle fences I make. In exchange, there would be some fencing repairs on the farm, firewood to chop and of course haymaking to help with that would cover my rent. It was a long-standing arrangement that suited both me and Tim and had never caused any tension. Though it would be several months before my return, I was already looking forward to it.

After the festivities and company of December, I was in a low mood when I pulled the wagon out on to the stone track beyond the yard, not helped by the rain relentlessly lashing the windscreen of the old wagon as it lumbered out between the elm hedgerows. I navigated the narrow lanes carefully, as they were still choked with the vehicles of residents and visitors who were drawing out the festive celebrations and had yet to return to work. I, however, was bound for Dorset and the market town of Bridport, where I would be working on a holding that hugged the Jurassic Coast.

JANUARY

The farm was called Frogmore. I had been offered a pitch on site for the duration of the contract, but the steep ground and narrow tracks that led to the buildings would make parking the wagon difficult and I had politely refused, arranging instead to stay at a friend's wood yard nearby. Here I could pitch up in the shelter of the sheds, under an old oak tree that grew from the hedgerow bordering the wide stone track that led to the yard. There were several primitive agricultural buildings. Two open-fronted sheds constructed from old telegraph poles and clad in corrugated-tin sheets were situated on the eastern side of the track, each one around twenty feet deep and slightly less than that in height, though both were higher at the front to allow access for vehicles. The bare chalk ground of the sheds had been covered by an array of mismatched pallets to keep the split logs off the damp earth and prevent them from rotting. Alongside the sheds stood an old road mender's living van in a state of disrepair; no longer suitable for habitation, it was being used to store bagged logs which would eventually be distributed to local retailers.

On the western side of the track were the cords of wood awaiting processing. Firewood is often still purchased by the 'cord', a well-stacked woodpile four feet high, eight feet wide and four feet deep. There were perhaps twenty cords or more of seasoned timber in the yard, waiting to be cut and split to replenish the store in the tin sheds opposite. Beyond the cord wood stood a more recently erected building, a long, low, timber-framed shed with a concrete floor and six bays that was being used to house the machinery needed for the work. An antiquated, rusting, blue Fordson tractor with a home-made hydraulic log splitter attached to its linkage was parked in one of the bays, sheltered from the weather. Tubs of oil and spanners of varying sizes lay strewn around an old truck in another; the truck was being used as a donor for the vehicle that was still used to deliver the firewood and was by now half-stripped of body panels and engine components. The end bay had been completely enclosed and served as a site office. It housed a wood-burning stove to fend off the winter weather and a small gas hob to prepare some warming soup or hot strong tea which could be enjoyed on the tattered sofa and armchair by the stove.

The oak under which I was parked lay just beyond the tin sheds to the rear of the yard. The stone track continued past the buildings, then petered out some two hundred yards distant as it entered a woodland, thick with goat willows and thorn scrub but scattered with clearings where bluebells and wild garlic would bloom in the spring. The ground under the oak was hard, quite suitable for parking the wagon, and my mood was lifted as I drank tea under its boughs while waiting for my younger nephew to arrive driving my pickup. He was yet to return to his contracting work in Bristol and I had promised him a day's wages to help me with the move. With my pickup at the yard I would be able to commute the short distance to Frogmore every day

JANUARY

and visit the nearest town for provisions. It was only shortly after midday when he arrived, and I had plenty of time to run him back across the border to Somerset and still be back at the wood yard before nightfall, which at this time of year was just a little after four o'clock in the afternoon.

The contract at Frogmore was a good eight hundred metres of planting work, and I had estimated it would take six weeks. I had been to look at the site the previous summer whilst staying at Home Farm, the West Dorset dairy where I had been based before the move to Eweleaze. It had been a glorious drive from Home Farm along the A35 trunk route that links the rural counties of Dorset and Devon, with exquisite views across Dorset's lush pastures to the coastline beyond. Frogmore was just off this road at the point where it narrowed to become a single lane before gently meandering into the village of Chideock. The farm's entrance was just a couple of hundred yards short of the village, across a cattle grid, and comprised a cottage and several outbuildings alongside some eighty acres of pasture that rose steeply away from the buildings to the coastal cliffs beyond. To the west lay the coastal hamlet of Seatown and beyond this, Golden Cap, the highest point on England's south coast that rises over six hundred feet above the Dorset beaches. To the north lay Colmer's Hill, perhaps the most recognisable feature of the West Dorset landscape. This hill, topped by a cluster of pine trees on otherwise featureless banks, rises steeply beyond the village of Symondsbury and stands clear above the surrounding countryside.

It had been a beautiful June day when I'd visited Frogmore to assess the work involved but the condition of the wind-battered hedges that still clung on here hinted at the harsh winter conditions I would be working in. The ancient hedge line had long since died back and was now just a broken row of declining hawthorns, their crowns swooping eastwards away from the prevailing wind. Thankfully, brambles were not a significant issue, although even on these weather-beaten slopes patches of them were trying to take hold and several of the remaining thorns had been engulfed in the briars' unrelenting grip. Once a hedgerow is in this condition, any thought of laying it needs to be abandoned. The hawthorns were too old and brittle; their gnarled and knotted trunks had long since lost the suppleness of youth and would likely snap if any attempt to pleach them was made. Regrowth, too, would likely be poor. Though young and well-maintained hawthorn will coppice and pleach well, as the tree ages it will tolerate this procedure less and it could even see the demise of the plant altogether. In this instance a decision was taken to leave the strongest specimens, remove the dead and dying, and plant in between using a native mixture of approximately sixty per cent hawthorn, interspersed with other species found within the

JANUARY

local, established hedgerows: dogwood, spindle, blackthorn, hazel, crab apple and field maple. Some dog rose would also be planted at intervals, as it is a welcome addition in any hedgerow.

The other problem at Frogmore was the exceptionally steep ground leading up to the site. I knew that when I returned in the wet season it would be impossible to drive to the hedge line, so all the plants and tools would have to be carried in a wheelbarrow from the closest vehicular access a quarter of a mile away. When I had looked at the work on a warm afternoon in June, I had thought how I rarely worked surrounded by such open country and beautiful scenery, though I was aware it was somewhat beguiling. Come January, when the inclement weather bore down, there would be little shelter. The twisted thorns, though beautiful in full leaf on a warm June afternoon, bore the signs of two hundred or more winters in their shape, hunched over with their backs to the sea, crippled by frost and gales. There would be no opportunity to light a comforting fire on these sea cliffs; it would instead be a test of endurance, out in the weather with only the gulls and rooks for company, digging the chalky topsoil to reinstate thorn and rose on the windswept hillside. With all this in mind, over tea on the lawn of the old cottage in the June sunshine, I agreed to take the contract and start as soon as Christmas was behind us.

It was just a little after eight in the morning on the first Monday of the New Year that I crossed the cattle grid to enter Frogmore. Predictably, the weather was coming in from the west and the wind, though not forecast to be particularly strong, was nonetheless buffeting the ash and pine trees that lined the approach to the farm buildings in the exposed coastal location. As I pulled up in front of the cottage, the illuminated kitchen window showed a snug interior, still decorated with festive lights. As the door to the cottage was opened, two dogs rushed out and I was greeted first by a black retriever and then his companion, a scruffy terrier of some description. Though both animals were excited, they immediately ceased their barking at a firm word from their owner, who followed them out. He was a Yorkshireman by the name of John Bell who had moved to this county with his wife and children several years previously.

My arrival had evidently been anticipated and shortly after John and I had begun discussing the upcoming work, John's wife, Elizabeth, joined us outside bearing a tray laden with bacon-and-egg sandwiches as well as a pot of tea. She placed the tray on the bonnet of my old truck and we talked as we ate, John explaining the most convenient route to take up the hillside while his dogs eyed him, and his breakfast, intently. Eventually he relented and threw each of them a morsel before scolding them, though they paid little notice to this and turned away for barely a moment before returning to their patient vigil.

JANUARY

John told me that despite the relatively dry December the hillside was by now unsuitable for vehicles, as I had suspected. After casting an eye into the bed of my truck and seeing the dented and rusting wheelbarrow I had brought from the woodsheds, he kindly offered me the use of a more suitable barrow to ferry my tools and provisions to the hedge line. With this, I should be able to get all the equipment I needed there in one load, which was some consolation; I estimated the journey, one way, would take at least a quarter of an hour up the steep slope and I didn't relish the idea of making it more than once – though I was told that should that be necessary, I was to be sure to knock on the cottage door for tea. Thanking John and his wife, I fetched the barrow from an old lean-to and loaded it up before starting the ascent.

The rough pasture of the hillside was populated with sheep belonging to a neighbouring farm. They had made trails that fanned out east and west from a weathered concrete water trough by the gateway before gradually disappearing out of sight over the hill's brow. A steady column of cloven hooves had turned these pathways into slippery channels of chalky mud that I was unable to navigate with the barrow. The grass, however, was not much better, as the fine rain that had begun to fall made finding a secure foot hold on the steep ground difficult. The keen west wind blew the rain into my face, and before reaching my destination I was forced to pull on a pair of waterproof trousers I had rolled up in my satchel. Although they were uncomfortable and tended to slip down unless secured by braces, they at least prevented my heavy corduroy trousers from becoming saturated before I started work.

Eventually, and to my great relief, after a quarter of an hour or so the slope began to lessen and the ancient hawthorns of the remnant hedgerow beyond appeared. They were a melancholy sight against the heavy leaden sky, their untidy crowns whipped mercilessly by an incessant, eroding westerly that blew unchallenged off the channel sea and forced them into unnatural shapes. Some of the thorn trees had fared better than others, but several of these veterans had already succumbed to the exposure and their sapless timber stood rigid against the wind, like weathered headstones. Their battle with salt air and gales lost, they now remained only as monuments to the farm's past.

Upon reaching the hedge line I took a moment to familiarise myself once more with the task before me. I crouched in a convenient hollow at the base of one of the old hawthorns, its spiralling trunk rubbed smooth by countless sheep who had taken shelter in this same place and who would likely be here now were it not for my presence. I poured a cup of strong tea from my flask and cupped it in both hands, which were reddened and cold after pushing the barrow from the buildings below. Feeling somewhat invigorated, I left the shelter of the hawthorn and headed for the western end of the hedge where I would begin my work.

JANUARY

The hedge ran parallel to the coastline, climbing gradually as it tracked east. The western end was the lowest point in the line, and it was here that the hedge required the most work. Stunted and chewed hawthorn, grazed by sheep, had become a tight nest of low scrub that spanned the gaps between more mature hawthorn trees. The few patches of bramble that had taken hold were more prevalent here at the lower end and the hedge line gradually thinned on its track east until it became no more than a hawthorn every thirty or forty paces with only a thick mat of rough grass between.

The eastern end would require little attention, except to knock the grass back to aid the planting of the young hedge. The grass would compete for moisture and nutrients and although the hedgerow would eventually win this battle, it would have to work hard from the outset of planting to gain the upper hand. In an ideal situation, any vegetation competing with the new hedge plants would be removed, or at least suppressed. On more accessible ground, the strip to be planted might be ploughed to remove this competition. Alternatively, a well-rotted mulch can be applied around the new plants that will not only suppress grass and any weed growth, but will also aid moisture retention, helping nurse the young plants through their first summer.

The juvenile plants here would have none of these advantages, and while I was in no doubt that they would eventually succeed, they would have to work harder and this would likely be apparent in their growth rate, meaning the hedgerow would take a season or two longer to mature. There might also be a few more plants that failed to take, which would need to be remedied by an inspection the following autumn and then made good by replacing any that had perished. Once the new hedgerow was established, around ten or twelve years after planting, it would be laid. This would invigorate the young plants that had succeeded in their battle against the grass and the elements and increase both their vigour and density, resulting in a healthy, robust and thick hedgerow. It would be the start of a new legacy for the hill and its occupants.

The hedger, like the forester, has to think of his work not as something completed in days or weeks or even months, but instead in years, decades and centuries. I thought on this often as I cut away the scrub and bramble to leave a clear line on which to reinstate the hedgerow. I was not the first man to engage in this work in this location, just the most recent in a line of countrymen who had shaped this timeless landscape and worked through its winters as I did now. Their presence could be seen in the hedgerows and meadows, copses and farms of this West Dorset landscape. I took some comfort in their ghostly company and despite the unrelenting wind and rain, which seemed to have settled in for the day, I was inspired to press on.

JANUARY

By the end of the first week at Frogmore I had succeeded in clearing the scrub from half of the hedge line. Every day I was welcomed by John Bell and his dogs. John would ask of my progress whilst his wife prepared me a hot breakfast which she would wrap in tinfoil to hold its warmth until I reached the hawthorn hollow on top of the hill where I would eat it looking out across the exposed chalk hillside. I warmed to the couple very much over the weeks, their generous spirit and good humour making the hard work easier to bear. Each day I worked methodically from west to east, cutting and raking the declining thorns and encroaching briars, clearing the line for the new hedgerow. Every twenty-five paces or so, less if suitable specimens were available, I was able to retain a veteran hawthorn that had the strength to grow on amongst the new plants. The poorer specimens, the dead and decaying, were cut back to make way for these new plants, and I stacked the rotten wood against the cut stumps to provide habitats for insects and invertebrates. The cut scrub was stacked neatly on the hillside, adjacent to the hedge line, and would be pushed up and burned the following spring once the drier weather meant the farm's tractor could make it up here. There was a good deal of firewood, too, that I had stacked in manageable lengths at intervals along the hedge line; this would be left to begin seasoning here on the hill before being brought down to the wood store once the weather and ground conditions improved later in the year.

The second Saturday of the month dawned wet and from the comfort of my bed I could hear the wind pushing through the oak boughs above and the constant drone of rain on the wagon's tin roof. It was gone eight before I rose and filled the kettle. While I waited for it to boil I opened the top of the stable door and took a breath of the fresh, early-January air. As I looked across the grass field that lay behind the long-shed opposite, a sudden fracas erupted from the rookery housed in a stand of lime trees that grew from an ancient hedge bank on the edge of the wood. A great number of the birds swarmed suddenly upwards from their resting place amongst the limes and were dispersed rapidly across the winter sky by the west wind which carried them at considerable speed toward my pitch under the oak. They made a great din as they glided overhead, quarrelling amongst themselves excitedly before turning into the wind and returning to the bare branches of the lime trees which swayed turbulently as the wind rushed through them. They seemed to settle as the kettle came to the boil, and I switched on the radio, waiting for the weather forecast for the week ahead. For a good part of the morning I listened to the radio and drank tea; after a week on the exposed hillside at Frogmore it was a pleasure to enjoy the warmth of the fire and have nowhere to be. The day was mine.

A little before midday I heard the distinctive rumbling of an old truck approaching up the stone track, which I recognised as belonging to my friend George Parsons who

JANUARY

ran the wood yard here. A Dorset native, George had cut timber and firewood from a number of woodlands in the district for decades and had established connections with many of the West Country estates. I had worked for him on several occasions over the years, both cutting in the woods and processing firewood in the yard, particularly around midwinter as the weeks prior to Christmas are traditionally the busiest for the firewood merchant. Though the majority of cottages are now primarily heated by gas or perhaps oil in more isolated locations, they will always have at least one fireplace, usually more. Many of these hearths remain unlit for most of the year, an open fire or woodstove requiring more attention than the more convenient gas, but as Christmas approaches there is a marked increase in firewood orders as people want to cosy up in front of a wood fire to enjoy the festivities.

I had not seen George in many months and he greeted me warmly after alighting from the old truck. He clasped my hand and enquired, 'How youm bin getting on?' I told him I was well, busy as ever, keeping my head down. George smiled approvingly. 'That's good to hear, Paul, there ain't too many that'll put up with biding all winter in the hedge, truth be told I wouldn't vancy it meself!' He said this with a broad smile, both of us knowing he'd spent more winters than he could remember hunched over a chainsaw, cutting and stacking timber in all weathers, and could relate as well as anybody to the testing physical work that hedging entailed and the endurance required to earn a living outside during the winter months.

We chatted briefly before George reached into the truck where his ageing longdog Whisky was resting awkwardly on the passenger seat. The old dog seemed put out by the cold draught that now entered the cab, his general air being one wholly intolerant of January's inclement weather. George drew from the cab a bag containing a crusty loaf and two tins of tomato soup. 'Let's get thek stove on in the cabin and warm us through, this bastard weather is in for the day,' he exclaimed. He put his tobacco tin in his shirt pocket and ushered old Whisky out of the truck. The dog took a lot of persuading to leave the comfort of his seat, but eventually tiptoed gingerly across the wet yard to the shelter of the cabin, where upon entering he made immediately for the tattered sofa. He climbed delicately into position before slumping down into the exposed foam of the cushions and promptly buried his nose into his middle in an attempt to retain as much warmth as possible.

George opened a string net filled with dry kindle wood whilst I filled the old kettle up from the outside tap and soon a good fire in the stove and hot tea in our hands were pushing back against the chill of the cabin. 'Youm still working alone is er, Paul?' George enquired whilst peeling the paper labels from the tins of soup and tugging back the lids by their ring pulls, just a quarter of an inch so they could warm gently on the hotplate of the stove. 'I am,' I replied, explaining that though I sometimes felt an extra pair of hands would be a help, this way I was only responsible for myself and had the freedom to come and go as I pleased without having to worry about the welfare of others. On occasion George employed the help of several woodcutters. The terms of their contracts were a piece rate, paid on the volume of wood they could cut and stack in a day, and so they were responsible for their own income regardless of weather or time off. George always made certain to set a fair rate to ensure that a competent woodcutter could earn a respectable wage. In this way George could accurately forecast the cost of converting standing timber into stacked lengths and therefore make an educated bid on any block of woodland coming to tender. He had many years of experience and a reputation for completing contracts in a timely fashion whilst taking great care to preserve the woodlands in which he worked. For this reason, Georgie was well respected by both the estate foresters who put the contracts to tender and the men he employed, and consequently was never short of either work or reliable, capable woodcutters.

'Tis finding the right sort of a bloke though in er, Paul, that's the trouble,' George went on to say. 'There's plenty what likes the sound of the work but then not so many likes the doin' of it, they don't makes 'em like they used to.' I agreed with him – in my experience it is a rare few who are prepared to tolerate such conditions and hardships.

In recent years many prefer to work from the comfort of a machine, or demand set start and finish times and an hour for dinner as well as time off during spells of particularly rough weather; it is a sense of entitlement that was unbeknown to me as a young apprentice and it is not the way of either hedger or woodsman. Work must carry on regardless of weather. There is rarely a day between October and March that isn't tainted by rain and mud, and when there is, when the weather swings round to the north or east, the biting cold, snow and sleet that it brings will soon make you crave the return of milder conditions from the wet and windy west. Of course, working in these conditions makes the onset of spring that little bit sweeter, the summertime a little more cherished. The rural labourer, like the birds and animals that reside in hedgerow and woodland, is undoubtedly more attuned to the subtle changes in the countryside as the year progresses.

George rolled himself a cigarette, a habit I'd given up years previously, though I still relished the distinct aroma as he puffed on the sweet dark tobacco. Steam began to rise from the soup tins that had been poaching on the woodstove and George threw me a crisp, dried leather glove from a shelf behind the stove so that I could hold my hot tin. He tore a hunk off the fresh white loaf with a calloused, greasy hand before passing it to me. Whisky perked up at this, gazing intently at my meal while George and I talked and ate. As it became apparent to the old lurcher that the loaf

JANUARY

was rapidly disappearing, he edged ever closer to me, raising a scruffy grey eyebrow and switching his gaze between the soup-soaked bread and me, in an effort to gain my attention. Eventually, in a last-ditch attempt to secure a morsel, he sneezed loudly in my direction, his face a picture of innocence as he licked his nose and muzzle apologetically. I looked at him, then George, who after a short pause roared with laughter. 'Thek old dog is a crafty bugger!' he exclaimed. I laughed too, and though I had been enjoying my dinner enormously, I couldn't help but admire the old dog's cunning and so I wiped out the tin with the remaining bread and gave Whisky the last, generous portion. The lurcher is a dog of much personality, always looking to exploit a situation to his favour and never working harder than he has to. A favoured breed of the Gypsy people, they have keen wits and an eye for an opportunity. I admire these qualities in both.

The second week at Frogmore passed much as the first. The weather remained largely unchanged and by putting in the best part of eight hours each day I had managed to clear the hedge line and was now happy that it was in a suitable condition to begin planting. The bramble had all been cut out of the western end, cropped short and raked clean. Piles of neatly stacked hawthorn wood were placed at intervals along the newly cleared line and further out, dragged well clear of the work site, were the piles of cut scrub and brash that awaited burning in the coming months. The thick mat of rough grass had been cropped as close as I could manage. This had taken me a full day, carefully setting up a string line at fifty-metre intervals and with the aid of a mechanical brush cutter, cutting a four-foot wide strip back to the bare, chalky earth. After a fortnight the preparation was now complete, leaving a clean, straight line cleared of vegetation and interspersed with mature hawthorn at roughly every twenty paces, ready and awaiting the new hedge whips.

It was a stark scene there on top of the windswept hill, the leafless thorns rising from the cold, bare earth. But as I sheltered in the hawthorn hollow, I allowed my mind to picture the scene in years to come. The new hedgerow would look beautiful here, weighted heavy with mayflower in the late spring as the swallows danced overhead. The spring lambs would find shelter beneath it from sun and wind, while the kestrel would hover in the coastal breeze above, taking voles and mice which strayed from cover. The owl, too, from a vantage point among the taller thorns would watch silently for these morsels on still summer evenings. The work here in January was hard going, but there was a great joy to be taken in this thought, and for a moment there under the old thorn, buffeted by the west wind, it was a calm warm day in May.

JANUARY

The following Monday I made the two-hour journey to the nursery to collect the hedging plants. Wherever I am working, I always source my plants from the same nursery which is located just a short ride from Bristol in the Chew Valley. I have used their plants for years and am always satisfied with the health and vigour of their stock; even in the most hostile conditions I have found few failures when returning to check on a new hedgerow's progress.

New hedgerow plants are generally purchased as two-year-old saplings. Bare rooted, and with a single stem just a foot to eighteen inches long, these young plants are known as whips because of their resemblance to this object. Nurseries specialising in trees propagate thousands of saplings every year before lifting them from the ground to order once the leaves drop and the plant becomes dormant, likely some time towards the end of November. Once lifted, the plants can be tied in bundles of twenty-five or fifty, depending on the size of their roots. Most have a single, searching tap root just a few inches long, though some, such as elder, may have a thicker mat of coarse, dense growth that will cling to the soil in which it was grown. Once lifted and bundled the whips can be placed in a bag, up to several hundred in number, and then tied securely; this keeps the dormant roots moist and away from the elements while leaving the tops of the stems exposed to the air. Stored in this way the young plants will survive for a week or more before they can be replanted in their permanent location.

If they cannot be planted within this time, they must be heeled in to protect their roots; this involves planting them in a temporary hole or trench in a convenient patch of earth, still in their tied bundles, until they are needed. They can survive like this for months, if needs be.

I had ordered the hedging plants for this work at the onset of autumn, as supplies are limited and a large planting scheme such as this requires a considerable number of whips. I intended them to go in at a density of five plants per metre, which meant I needed nearly four thousand whips. As always, the plants looked healthy and robust upon collection, and between the whips, the bamboo canes for support and the spiral rabbit guards my truck was fully laden as I made the return journey to the woodyard. When spending time in the countryside you may notice these sections of new hedgerow, the plants wrapped in plastic or set in tubes to protect them in their early stages of growth. There is no doubt that these measures do afford the young whips protection; rabbits relish the soft young plants and can make short work of a newly planted hedge line. They are, however, not always necessary, and when embarking on a new planting project I will always enquire about the rabbit population before resorting to guards as, though effective, they do nothing to enhance the aesthetics of the countryside. They are rarely collected once the new plants are established – more often they are left to become brittle and break up at the base of the hedgerow. If there is not a problem with

JANUARY

rabbits in the area, these guards should be avoided, as the young thorns and other hedgerow shrubs are hardy and quite capable of looking after themselves. After all, hundreds of thousands of miles of hedgerows were planted on these isles without any need for plastic guards or the bamboo canes which support them.

In my opinion a longdog, a satchel full of purse nets and a box of ferrets is still the best way to control an expanding rabbit population; in this way they provide a good meal for the warrener and his dogs and a healthy population is maintained. I find little more sickening than the sight of a rabbit ravaged by myxomatosis, a brutal disease introduced to control their numbers many decades ago, and which is still prevalent here in the West Country. The warrener, however, is becoming a rarity, and with rabbits in abundance on the hillside at Frogmore and no viable alternative, the whips here would have to be protected by plastic guards.

The planting of a new hedgerow is straightforward. I generally manage two hundred plants a day with the appropriate guarding and so two to three hundred metres per week is a realistic target, more if conditions are good. Here on the hill, the weather continued grey and wet, but for the current task this was preferable to colder, more settled conditions. A hard frost would make the earth unworkable and even if, with great effort, holes of a suitable depth were able to be dug, the frozen chalk would burn and damage the young roots and the plants would likely perish. Soft, easily workable ground might see tools and boots heavy with clotted mud, but it provides a suitable bed in which the roots can settle. The coming months would see bitterly cold days and hard frost before the warming days of April, but with their dormant roots bedded in when the earth was soft and forgiving the new plants would be quite safe, and the constant rain from the west would soothe their transition to the chalk soil of this Dorset hillside.

My days settled into a steady rhythm. The whips had been lightly heeled into the vegetable patch which lay just a short distance beyond the cottage. John's wife, Elizabeth, would prepare a hot breakfast roll whilst I lifted the plants I needed for that day, placed them in a bag and tied it securely to prevent the wind drying out the roots before they were transplanted. These went into the wheelbarrow alongside a bundle of bamboo canes, the spiral guards and my satchel of provisions. As when laying a hedge, it helps to have the workload for the day set out before me; I would be finished when all the plants I'd loaded into the barrow were in the ground.

The day always started with tea and breakfast in the hollow beneath the old hawthorn. This was particularly welcome after the steep climb from the farm buildings below. The rooks, clever birds, learned quickly that a morsel

JANUARY

of bacon rind or perhaps a yolk-soaked crust could be extracted from me on some mornings, and several often gathered enquiringly just a few yards from where I sheltered. Tattered and scruffy, they seemed not to mind the exposed location, and I couldn't help but admire their resilience.

I made my way from west to east, digging the holes ready to accept the new plants. Having first pulled a string line tightly along the length to be planted to ensure the new hedgerow was straight, I used a length of hazel, cut at a metre and marked halfway by a score in the bark, to evenly space the holes. I have found that the best tool for planting is a heavy, spade-ended fencing bar; the spade end split the ground easily, and I eased it back and forth to open up the earth ready to accept the roots. Working along the line, once I had reached the target length for that day I turned and worked back on myself. Discarding the hazel measuring stick, I went back along the row, putting another line of holes parallel to the first line but at approximately a foot distant, resulting in a double staggered row.

I lifted the fifty-strong bundles of thorn from the sack, then mixed two and a half bundles of thorn with around seventy-five whips of other species. This provided a random assortment but with hawthorn the dominant plant, as this was the species that would make certain of the new hedgerow's stock-proofing capabilities. The plants were now ready to go in. It is worth taking the time to dig the holes for them correctly; the new plants should drop easily into position without having to be forced, which could see unnecessary damage to the roots. Ideally the whips will settle easily into position and can then be gently raised upwards to make sure the roots are not bunched or crowded. On close inspection of a young hedging plant, a collar is apparent toward the top of the root, a slight discolouration that shows the depth at which it was planted at the nursery; the whip must be planted to this same depth in the new location. Finally, a boot heel is brought down around the whip to ensure the earth holds it firmly in position; that good contact between the earth and roots will prevent any of them drying out.

This procedure took most of the morning, often longer, and would become uncomfortable by the afternoon, as I'd been first working with the heavy bar and then stooped over, heeling in the new plants. Applying the canes and guards was lighter work. Putting a cane adjacent to each plant, though being careful to avoid the roots, I tapped each one with a club hammer to ensure it sat straight and wouldn't fall in the wind. Finally, the plastic spirals were put into place. On smoother species, such as crab apple and spindle, these can simply be dropped over the cane and plant, but on hawthorn and blackthorns that have horizontal growth even at this early stage, they have to be unravelled and carefully wound around the young plants.

JANUARY

By the end of January, I had completed a good length of the planting contract at Frogmore. Despite persistent wet weather and gales, I had made steady progress, encouraged by the kindness and appreciation of John Bell and his wife, Elizabeth, and I felt pleased to be a part of their vision for the farm and its land. On the last Friday of the month I was driving back from the farm to my pitch at the woodyard in a relentless westerly wind. As I pulled out on to the main carriageway the headlights of the oncoming vehicles glared and blurred in the drizzle. I made my way eastwards, up the hill that led away from the village, and there was enough light remaining to make out the line of the new hedgerow. As I had expected, the work here had been testing, though despite the difficult conditions the month had seemed to pass quickly and I knew the contract would be finished in good time. I felt a great sense of satisfaction that as my work on the hill neared its end, I had become a part of the land's story here.

7

February

I had been awake for a while, lying in the darkness, when I was made aware of the approaching morning by a song thrush which had recently taken up residence in the oak branches above the wagon and was now loudly proclaiming it his territory. His arrival had brought with it an awareness of the progressing year, and that I still had much to do before the season turned. The subtle signs of change had started to creep in, even on mornings such as this, when a light frost covered the meadow behind the woodshed and a thin layer of cat ice locked the muddy water within the ruts of the track that led to the wood.

The rooks in the lime trees were also busy. They had started building last month, renovating their roost high in the limes, which had been battered by the worst of the winter rain and gales. Even at this early hour, the trees barely silhouetted against the lightening February sky, the rookery was crowded and cacophonous. It reminded me of an avian equivalent of a scene from a Dickens novel: a nest of thieves and undesirables, the corvids' voices were rasping and coarse and each bird was deeply mistrustful of his neighbour, who was quite likely to take any available opportunity to steal an appealing twig or stick for his own abode should his companion's attention be distracted for even the slightest moment. Now, in the still winter's morning, their raucous quarrelling carried for miles across the countryside and was in stark contrast to the melodious notes of the thrush above me.

The work at Frogmore was completed and the westerly wind and rain that had accompanied me on that contract now abated. As the sun gained height behind the oak there was hardly a breath of wind. The frost that had carpeted the meadow at dawn soon thawed and left the pasture glassy as the sunlight reflected off the moisture that still clung to the grass; its bright rays held little warmth and were as yet unable to dry the ground. The ditches that accompanied the ancient hedge banks around the yard were now full of standing water. The low-lying ground to the south was a network of pools, teeming with gulls which had made the short journey inland to feast on the worms and grubs that floundered in the water-logged earth. Several rooks skulked amongst the gulls, ever-present opportunists hopeful of snaring an overlooked morsel between their nest-building forays.

It was a bright, late winter's morning and although the

air was cold it was a pleasure to be outside. After a breakfast of eggs and toast I decided to walk the short distance up the track that led through the yard and onwards to the woodland at the rear of the site. The wood here was unmanaged, an abandoned corner of the estate, and contained mature oak, lime and ash. Beneath this grew an understorey of hazel, thorn and holly, battling against the goat willow that had become established here and now formed dense, twisted thickets in the woodland's wettest areas. The willow takes root where other woodland species are less adept, in the low, muddied hollows that remain waterlogged for much of the year. The tree grows quickly and soon out-competes any other vegetation, putting down roots wherever the stems from the pioneering tree contact the swampy earth. But while it might root readily, the goat willow will never make a substantial tree, as once the stems become heavy with age its soft timber is unable to support its increasing weight and so will collapse. The fallen stems, however, whether completely severed from the main plant or still partially attached, will propagate and go on to establish a new thicket. The goat willow exhibits the same behaviour in hedgerows and spreads aggressively wherever it is able to gain a foothold. It is often seen on West Country hedge banks that have been left unmaintained and can render great lengths devoid of all other species where the willow has bullied its way to prominence.

As I made my way through the wood, the sun gained

height and the trees were illuminated by the soft February light. The woodland, like the pasture, was saturated and the constant sound of dripping water accompanied me as I walked. A sudden shower of droplets descended from the ash trees a short distance from the track, alerting me to a squirrel which navigated through the understorey, twitching his tail nervously as he leaped between the mocks of hazel. The grey squirrel and the woodsman have little affection for each other. Though his gymnastic skill and his work ethic are to be admired, if left unchecked the grey squirrel can cause huge amounts of damage to a woodland, and swiftly loses any appeal to a forester or woodcutter working a plantation that has been decimated by his attentions. An opportunistic feeder, the grey squirrel will strip bark and buds, preferring younger specimens rich in nutrients, and his activities can kill or deform a great number of young woodland trees. He will work tirelessly through stands of beech and sycamore, which he prefers above all others, though young oak and sweet chestnut are also vulnerable. It is perhaps some small comfort that he has little appetite for the ash, as this native tree, once so familiar whether open-grown in our hedgerows or forming close-knit stands in ancient and commercial woodland, is already fighting on another front against the virulent 'ash dieback'. During the spring the grey squirrel will expand his menu to include fledgling songbirds which he will hunt in scrub and copse, greatly reducing their survival rate. Then of course there is

FEBRUARY

his impact on the native red squirrel; having out-competed the red for food and habitat, the grey then sealed the reds' fate with a pox to which he was immune while the reds had no protection. On this day, however, I enjoyed the moment, watching him dance among the swelling crimson buds of the elder and the lengthening catkins on the hazel, while the soft light illuminated the damp woodland.

The hedge bank that marked the eastern boundary of the wood was still a prominent feature, standing four feet clear of the track on which I now walked, and topped by large clumps of hazel, or mocks as Dorset woodsmen refer to them. The absence of livestock had preserved the bank to a large extent, though the woodland was frequented by fallow deer, and several deep cuts along the length of the bank clearly showed the route they took to access the pasture lying just beyond. The ditch on the far side of the bank, which was now full of water, was easily navigated by these nimble animals. On this side of the bank, great swathes of snowdrops grew in the dappled shade beneath the hazels, and pockets of them were visible here and there beneath the larger trees of the woodland to my west. For me, the snowdrop is perhaps the most beautiful of our woodland flowers. Its simplicity and understated elegance light up the dormant woodland during the late winter and bring a promise to all who are becoming weary of the short winter days and grey February twilight: spring will come.

Days like this are a gift and offer respite from the grip

of winter, if only for a brief spell. All of us here – the rooks in the limes, the fallow in the wood, the songbirds in the trees and hedgerows, the squirrel dancing in the branches, the frogs whose spawn gathered in great bunches in the ditches and dew ponds and whose low croaking now resonated gently – were aware of the progressing year.

On returning to the wagon, I decided to light a fire and cook my evening meal outside. In my opinion, there is nothing better to lift the mood than the company of a good fire, whatever the time of year or weather conditions. There had been a lot of standing deadwood in the hazel mocks on the hedge bank which was easily snapped off, and I had returned from the wood with a sizeable bundle of dry firewood. Within minutes a decent fire was burning. The sun was by now sitting just above the lime trees and the temperature, which had been pleasantly comfortable, was beginning to drop. I pulled on my woollen smock before thumping the kettle iron into place with the club hammer and hanging the kettle amongst the flames. The flames licked the base of the kettle, curling around its blackened spout and causing it to pop as the water within warmed and began to hiss. The smoke from the fire rose steadily in the windless air, an occasional waft bringing it in my direction and forcing me to turn and hold my breath. Once the kettle

FEBRUARY

had come to the boil, I added two good lumps of oak on to the bright hazel coals, allowing them to catch before adding milk to my tea and reclining on the steps of the wagon. Though a strong flame boils a kettle in good time, it is of little use when cooking, as the intense heat of the flames will cause the food to burn and stick to the base of the cooking vessel. For cooking you want to let the flames die down and use the hot coals. Oak coals are best suited for the task, as the dense wood provides slow, consistent heat. I had drunk two cups of tea before the oak had burned down enough to begin cooking, and the embers flared red hot as I knocked them about with the hook of the kettle iron and prepared a suitable bed on which to place my trivet.

    I enjoy putting together a proper stew, when I have the time and inclination, and here underneath the oak in the mid-afternoon sun, I set to making a simple sausage stew. I nestled the pan in the hot ash next to the glowing coals and browned off the sausages with some onions and a good knob of butter. Then I threw in a few rashers of bacon for good measure, before adding a couple of tins of tomatoes, some chunks of potato and whatever vegetables I had in the wagon's larder. Finally, I added a bottle of ale I'd been saving for this occasion. The old saucepan was now half full of simmering broth. I placed it on to the trivet and put the lid on, allowing a small gap for the steam to escape, this would keep out the majority of ash that became airborne and rose in the heat. It was now a waiting game while the stew gently boiled down and thickened, with perhaps three quarters of an hour or so of regular stirring to ensure the whole mixture was thoroughly cooked through. I was in no rush, however, and content to drink the remaining ale on the wagon steps whilst watching the reddening sun sink slowly behind the rookery.

It was light by the time I woke the following morning. I had slept soundly under the boughs of the oak after consuming a generous helping of stew and several bottles of ale, and even the thrush's familiar song had not disturbed me. Upon opening the door of the wagon, I was greeted by a bright, clear day, much like the previous one. I lit the stove and as I waited for the kettle to boil I listened to the by now familiar sounds of the wood yard and the distant ringing of church bells pealing out across the Dorset pastures from the village several miles away.

With the work at Frogmore completed, I had just one more contract to fulfil before the hedging season would be over for another year. This final contract consisted of two lengths of hedgerow to be laid, one of hazel that ran to a hundred metres and a shorter length that consisted primarily of elm. Both would be laid in the Devon style and were located in that county, just a little over an hour's drive from

FEBRUARY

here. I took my time packing up the wagon, leisurely drinking tea beneath the oak and not at all in a mood for hurrying, taking in the sounds of the yard and fields about me, enjoying the sunlight as it reflected off the dewy meadow, being serenaded all the while by a robin who flitted amongst the hedgerow behind me. It wouldn't be long before the birds were pairing up, I thought to myself, and the robin would likely be nest building within a month.

Eventually I manoeuvred the old wagon gently from the pitch under the oak, carefully navigating around the woodsheds before turning on to the stone track that would lead me out to the lanes. I had spoken with George and already said my goodbyes, though before pulling away I made sure to leave a bottle of the Somerset cider he was fond of and scribble a quick note of thanks. He was an old friend, and I was grateful for him accommodating me over the last few weeks. My pickup truck would remain here at the yard while I finished the season in Devon, and I would arrange collection when I returned to Marsh Farm in March. For now, it would be secure here, tucked away at the end of the old tin log-shed.

Though there was a distinct February chill, the weather remained calm and bright and the blue skies were barely tainted by the occasional light cloud that held no threat of rain. I was comfortable in a woollen jumper and cap with the cab window open. Already the days were growing noticeably longer and I was aware of the occasional green hue to the passing hedgerows as the leaf buds began to swell on the hawthorn. Winter's grip on the country remained, of that there was no doubt, but here near the Devon border there was the slightest hint of the approaching spring. As I made my way along the lanes I saw that work on the farms was gaining momentum. Over the hedgerows a tractor was spreading slurry on one of the drier fields to encourage a good crop of spring grass. The dairy farmers would already be looking ahead to warmer days and would be hoping to take their first cut of silage in April. In a vast wheat field on the other side of the lane, another tractor straddled the tramlines and applied fertiliser to the crop that had been dormant over the coldest months. Now the lengthening days coupled with the fertiliser would see it begin growing in earnest until its eventual harvest during the high summer.

I stopped briefly in Bridport for provisions before heading south-west on the main road. I passed Frogmore and cast an eye south over the completed contract. The remaining old hawthorns that dotted the new hedge line were clearly visible from the road, although unnoticed by the passing traffic. These thorn trees had watched over this road for centuries and I couldn't help but give them a nod to acknowledge their unwavering vigil as I passed. Presently, I approached the hill that marked the western end of the village of Chideock, where the road climbs steeply at a formidable gradient and leads eventually into the village of Morcombelake and the Devon border beyond. The climb may be barely noticeable in a modern vehicle but the old wagon, like a tired carthorse lacking the enthusiasm and stamina of its younger years, must be gently encouraged to make the ascent and not hurried. Fortunately, from its outset, the road becomes two lanes which allowed the growing queue behind me to overtake and carry on their journeys unhindered, as by the time we reached the brow of the hill we were somewhat off the pace at a little more than five miles per hour. But we had made it. The faithful old motor has never let me down thus far.

The road led on across the border into Devon, and within an hour I had passed through the market town of Honiton and crossed the River Exe, passing the city of Exeter. I left the main roads and began negotiating my way through the narrow Devon lanes and deep valleys that led to my destination at a holding called Culme Farm. Despite the clear weather, the lanes here were wet and the adjacent fields remained waterlogged from the heavy rain that had persisted after Christmas. This caused a constant run-off as the fields now drained into the narrow, potholed Devon lanes. The roads were bordered on each side by steep banks that left little room for passing traffic and the hedgerows on top of the vast majority of these were in varying stages of decline. Some of the banks were devoid of vegetation altogether other than rough grass or perhaps a little bracken and bramble alongside the occasional hazel mock which inevitably showed the scars of mechanical trimming. Others had been spared the rotary flail and the hazels now grew tall, leaning into one another from either side of the lane to form a tunnel of branches. They whipped and scratched the wagon's black-painted exterior as I passed beneath and I had to pull over to make sure I had retained the witch's hat which perched precariously on the chimney to keep the rain from entering the stovepipe. I feared that the thin band of steel and rusting bolt that held it in place wouldn't withstand the grasping branches. Although the hazel trees were still devoid of leaves, each branch was now decorated by the golden catkins often referred to as lamb's tails by children in the countryside, whose appearance coincides with the early lambing on many of the farms.

The lane widened as I drew closer to Culme Farm and the banks were now lined with beech. In the south-west corner of England, beech has been widely used as an agricultural hedging plant and is seen across the region's uplands, most noticeably on the farmland of Exmoor where great lengths are still regularly maintained by laying alongside mechanical trimming. Other lengths of beech hedgerow, like the one I currently observed, have been long forgotten and the mature trees now line the quiet lanes of Devon's backwaters, ancient avenues of remnant hedgerows that, though no longer suitable as a barrier to livestock, still enhance the beauty and character of these wild and little-known hillsides on the edge of the moors.

The road levelled off and once again the vegetation on

FEBRUARY

the banks became sparse. A wide vista of unbroken farmland unfolded before me and in the bright afternoon light I was clearly able to make out Dartmoor away to the south. After less than a mile I came to a gap in the hedge bank and a cattle grid, next to which was a heavy slab of slate roughly hand-painted with the name 'Culme Farm', and an arrow pointing down the hill to where the farmhouse lay at the bottom of the valley. I descended the steep drive, noting thankfully that it was well made up with stone at the top and concrete nearer the farmhouse, meaning that I would have little problem when the time came to make the return journey, something I always have to take into consideration when locating the wagon in some out-of-the-way pitch.

Like both Eweleaze and Windwhistle, Culme Farm had until recently been a small working dairy. It had been outmanoeuvred by the bigger intensive operations in the neighbouring valleys and so the farm had closed and its original inhabitants were now gone. As I looked at the beautiful countryside surrounding me, I thought what a wrench it must have been to leave this place. I felt for the family that had farmed here, and no doubt fought hard to keep the business afloat. The demise of small family-run farms is now unfortunately a familiar story all over our countryside. The new owner was a man by the name of

Jim Keen, a successful businessman who was originally from the Midlands and now found himself fortunate enough to retire from city life to this quiet valley. Jim and his wife, Katherine, had been here for eighteen months and had already carried out extensive work on the house and outbuildings, converting cowsheds to stables and installing an arena in the old farmyard for the horses that Katherine kept. Jim informed me that she had upwards of fifteen of the animals and that they employed a full-time groom to help with their upkeep and exercise. Jim's attention was now turning to the management of his newly acquired land, and alongside the hedging work I was to carry out, a party of men would be working on the drystone walls on the property and an agricultural fencing outfit would be installing several miles of post and rail to secure the horses.

As I looked across the scene, I couldn't help feeling a sense of loss. The work Jim was putting in place would be of great benefit to the land here, of that there was no doubt. The restoration of the hedgerows and walls would improve habitat and biodiversity, and he intended to manage and restore the woodlands here too, along with the streams, pools and adjacent wetland. His vision for the site was to be commended. But my thoughts couldn't help lingering with the previous owners, of whom I knew nothing beyond the little that remained here to tell of their life in the valley. These old farming families are tied to their environment in a way that, perhaps, it is hard for

FEBRUARY

a newcomer to understand. Their presence here often spans many generations, and more than just living on the land, they are a part of it, the very fabric of the countryside deeply intertwined with all aspects of their life and livelihood. And so even as Jim outlined his plans, gesturing enthusiastically up and down the valley, pointing out upcoming projects and ideas, I couldn't help but think that although the valley had much to gain, the closure of the farm had been a great loss. With the original, indigenous custodians now gone, their knowledge and way of life were lost from this isolated Devonshire valley for ever.

There was now just a week left of February. The contract here would take me through until mid-March and would be my last of the season. By this time of year I was beginning to feel a little weary of the work; the good weather of late had left me craving the longer days and warmer evenings of the season to come, and as I prepared tea by candlelight in the wagon on the first evening I struggled to muster my enthusiasm for this final push. The hedgerows to be worked on at Culme linked a small copse to the larger woodland that rose from a clear pool at the valley's base to cover three or four acres of hillside. The woodland was made up predominantly of ash and oak with a good understorey of hazel and holly. The hedgerow was made up almost entirely of hazel and sat upon a steep bank, as is usual for Devon and the neighbouring counties. There was bramble and elder here, too, although it wasn't excessive, and notably both hawthorn and blackthorn were completely absent from the boundary.

The bank itself was in good condition and patches of gorse scrub grew along its steep face, the vibrant yellow flowers already in bloom. The gorse first flowers in January when its delicate gold leaves brighten the winter twilight and provide an early source of pollen for butterflies and bees, and it will continue flowering throughout the spring and well into the month of June. It was said by country people that when the gorse was in flower, it was the season for kissing, though with several species of gorse common on our isles all flowering at different times throughout the year, there is rarely a week that these lively little blooms are not visible, no doubt something the country folk who coined the phrase were well aware of.

The hedge was only a short walk from the wagon and was clearly visible from the window above the desk, so my progress each day could be easily surveyed from the comfort of the wagon's interior. Despite being pitched so close to the hedge, I intended to maintain my habit of packing tea and provisions to last the day before setting off each morning. I have found that my most productive work is carried out before lunch, and I will complete around two thirds of the day's workload during this time.

FEBRUARY

After enjoying an outdoor lunch I will then build on the morning's achievements and conclude that day's stretch. If I enter the warm confines of the wagon before the day's work is done, though tempting, it will then require a huge amount of effort to leave its comfort and return to the often unwelcoming or even hostile weather that can accompany me during my winter work. The comforts of the woodstove, fresh hot tea and the old armchair are best left until the day's work is finished, and the mere thought of the relief they will provide from the biting cold or driving rain is invariably enough incentive to complete the day's task.

Though the year was still young, I awoke to a dawn chorus of considerable volume. The woods were just a few hundred yards from my pitch and the trees were already busy with birds laying claim to nesting territories. The robin, my constant winter companion, was easily distinguishable; he is an early riser and is usually in full voice at the first hint of the morning light. The song thrush, too, was awake and his melodic tune resonated around the valley; the steep banks and isolated location stifled other familiar countryside sounds and acted as an amphitheatre to amplify his voice. As the light increased, the distinctive, piercing song of the wren reached me at the wagon; he rose later than the robin and song thrush, but was forgiven his time-keeping for the enthusiasm with which he serenaded the new day.

I packed my satchel for the day ahead before pulling on my boots and woolly hat and, lastly, gathering my tools and fuel from the locker beneath the old wagon and then making for the copse on the hill where I would start my work. The weather ahead looked promising with just a little broken cloud; there was no frost on the ground and it felt mild despite the clear conditions. By the time I reached the hedgerow I had already removed my woolly hat and stuffed it into the deep pocket of the torn, waxed jacket I wore over my shirt. Surveying the hedgerow, the job looked straightforward. The Devon style I would be working in doesn't use stakes or binders, and I could see that the hazel here contained plenty of the 'crooks' that I would need to peg the laid stems firmly to the top of the bank. As the blue tits and wood pigeons belatedly added their voices to the chorus and the first rays of the sun climbed above the hill and illuminated the tops of the oaks in the wood, I felt uplifted, my enthusiasm for the work reignited.

Hazel hedgerows are commonplace in the south-west and are often planted when the hedge is being used to mark a boundary, such as a parish boundary, rather than for the retention of livestock. The hazel is one of the easiest and most forgiving plants to work with, especially if it has been maintained in a proper fashion and the stems are of a

FEBRUARY

suitable size, ideally no larger than a man's arm in diameter. While it is possible to lay larger stems, the hazel's habit of throwing up multiple rods from its stool, which tend to fuse together as the plant ages, can make this very difficult. When laying in the Devon style it is of paramount importance to bring the stems in close contact with the ground, and hazel that is 'over-stood' and has become tall and heavy will have fused together right at the base, making pleaching them at the correct height virtually impossible. If coming across this situation the best option is to coppice the hazel and then lay once suitable new growth is available.

Here at Culme Farm the hazel was in prime condition for laying and I estimated had been coppiced some twenty years previously. Hedgerows have always been exploited for timber, firewood and other countryside produce, and a hazel hedgerow will produce a good harvest of rods that will have any number of uses around the farm. The rods gleaned from this one would probably have provided bean poles and pea sticks for the kitchen garden of the farmhouse; from up here on the hill the raised beds were clearly visible and had obviously been in use until the present owners took over, providing a glimpse into life in the valley when the dairy was still working.

With it not yet being nine o'clock, I began my work at the eastern end of the hedgerow, where it entered a small copse that itself contained a good deal of hazel alongside a stand of several oaks and an abundance of elder. Elder frequents broken ground and its presence can often indicate some kind of earthworks where recent disturbance has exposed clean soil for the plant to establish itself. Further investigation into the spinney showed a badger sett that was quite obviously still active; the fresh earth at the different entrances and the smell of their latrine further back in the woods a clear sign of current activity.

I began by coppicing the first ten feet or so and raking the top of the bank clean for the first stems to be laid upon. I then proceeded to work down the row, thinning out the hazel to leave only the stems I intended pleaching and laying, stacking the brushwood tidily on the lower side of the bank, and all the while separating any rods that had a convenient fork which could be used to peg the whole structure in place at the end of the day. I was in no rush with this hedgerow; the previous contract at Frogmore had been completed in good time and when planning my season in the autumn, I had thought that it might already be March before I was able to get here. As it was, by the time I stopped for tea shortly after midday I had already prepared the first twenty-metre length for laying and the easily workable hazel coupled with the fine, dry weather made it a pleasure to be out working on the hillside. The birdsong had lessened by now, though still I was accompanied by the robin who sang from the hazel on the edge of the copse, and at intervals dropped into the newly thinned hedge line to search amongst the stools for grubs that had been displaced by the morning's activities.

FEBRUARY

I worked methodically through the afternoon, pleaching the smooth rods in turn and bringing them down gently into place on the bank. There is a great satisfaction to be had when laying a hazel hedge; the long straight rods provide a consistent and tidy structure and the straight-grained wood splits easily when the pleaching cut is made. With care, a real showpiece can be made, the consistency and uniformity of the hazel working with and accentuating the hedgelayer's craft. After dealing with thorn trees for most of the season, which bit and scratched as they were brought to heel, the hazel felt cooperative and required only the gentlest persuasion to fall into line. As I worked, I took great care to maintain both the width and height of the emerging structure and the final form sat at about two feet in height and perhaps double that in width. The bank itself was approaching four feet high and was steep sided which meant the whole structure was approaching six feet in total and formed a formidable barrier. It served well as a boundary, and the bank's sound condition was a credit to the previous custodians here who had obviously taken pride in its maintenance.

With the pleaching now complete I rested once more and poured the last of the tea from my flask. As I cast an eye south-west down the valley, I could see in the distance the wild landscape of Dartmoor which rose from the surrounding, tamer-looking pastures. Though I have frequented many of the West Country uplands, laying hedges on both

Exmoor and the Quantock Hills and cutting firewood on the Mendips and Blackdowns, as well as enjoying many an early morning in the late spring stalking the trout which rise for the hawthorn flies on the upland reservoirs of Bodmin Moor, I have spent little time on Dartmoor. As I looked down the valley upon the scene, I could see the irregular shapes of broad-leaved woodland coupled with the sharper lines of more recently planted conifer plantations, between them largely shielding the Dartmoor town of Okehampton from view. This town, like many in the west, prospered off the back of the medieval wool trade, and is now growing once more. Devon is a rural county in which agriculture and land management once kept a good deal of the local population employed, but there is now little need for this labour force, and where work can be found, the wages in agriculture are amongst the lowest in the country, barely enough to sustain a young family and almost certainly not in the desirable villages and hamlets nestled amongst the Devonshire farmland. Unable to afford the considerable cost of living in the villages where they might have family ties stretching back generations, many people are being forced to move instead to the peripheries of the old towns and cities, such as Okehampton, and seeking employment away from the land. This is why I find myself living as I do, travelling the lanes and byways of the agricultural landscape of these south-western counties and plying my craft. Urban living

FEBRUARY

is not for me, and I find myself uneasy away from the hedgerows and woodlands.

My next task was to cut the crooks from the hazel rods that I had put aside earlier in the day. These were rods that had a convenient side shoot of a sufficient diameter to make a hook or peg to keep the newly laid pleachers in place. I had cut these crooks to roughly three and a half feet before using my axe to sharpen the end to be driven into the ground, leaving about six inches or so at the other end clear of the side shoot that formed the hook. In this way I would be able to drive the crook into place firmly with a sledge and then cut the stub end off neatly above the hook to leave a clean finish. As I had laid the hazel stems, I had tucked them into themselves as much as possible and this now held them in the desired position, for the most part. The crooks would be driven in down each side at regular intervals to secure the edge of the newly laid hedgerow and firm up the whole structure. As with stakes and binders on the Midland hedge, crooks are designed to hold everything in place while the hedge is at its most vulnerable until the new growth comes through and strengthens the barrier.

I worked along the lower side of the hedgerow first. At every pace I drove in a crook until the hooked top clasped the pleached stems tightly and held them rigidly in place, keeping them in as straight a line as I was able. When I reached the point where I had finished the day's cutting, I stepped over the hedge and worked back along the top side, driving in the crooks at intervals so that the upper and lower lines were not directly opposite each other but instead staggered, ensuring as much rigidity in the structure as possible. As the daylight began to wane and the temperature dropped, the birdsong once more became the prominent sound in the valley. I busied myself with pruning saw and secateurs, trimming back the hammered stubs from the crooks and tidying any stray branches that ruined the line. I felt keen to leave my mark on this ancient hedgerow, and was pleased with the day's work.

By the time I collected my satchel and filled the old feed sack with dry hazel wood for the stove, the first stars were beginning to appear, and as I made my way down the valley and back toward the comfort of the old wagon, I became aware of the night-time chill drawing in. Away in the woodland the tawny owl called from the treetops. These birds would soon be nesting, rearing their brood in some convenient hollow and patrolling the hedgerows and woodland by night for mice and voles. Their chicks would likely be on the wing by the time the hawthorn blossomed in May.

The weather remained calm for the final week of February and by the time the sun gained height and its rays reached the valley floor I was usually able to comfortably dispense with my woollens and work in short sleeves. Even then, the warmth of the sun would often cause me to break sweat. By the time I stopped for tea in the mid-morning, the wood pigeon would be cooing from a thick stem of ivy that clung to one of the oaks in the copse. As the daylight lengthened and warmed and the birds became more vocal, it was easy to be lulled into believing that the winter was fading, but although the countryside was rousing from its dormancy, the year was still young and winter was not yet over.

It is at this time of year that the hedger must pay close attention to the habits of the birds that nest in the hedgerows, as the end of the hedgelaying season is governed by their activities at this point. The mechanical trimming of hedgerows comes to an end at the close of February. It is all but impossible to identify nesting sites when enclosed in a tractor cab and deafened by the din of the rotary flail, and so although it has its place in hedgerow maintenance, the flail needs to be packed away at the earliest onset of spring. The hedgelayer, working in closer, quieter contact with the hedgerow, can survey the lengths on which he works before starting each day, paying especially close attention to thickets of bramble and ivy, which have particular appeal for farmland songbirds. In this way, he may continue his work into March and perhaps even April if the spring is late and the birds are reluctant to begin nesting. This, however, is rare and generally by the second or third week of March the hedgerows must once again be left by the hands of men and returned to the birds. The robin and wren, the thrushes and finches, the blue tits and blackcaps, whitethroats and yellowhammers, all these and more will find safe haven within the hedgerow.

My season had gone well and was coming to an end in good time. With the hazel hedge complete there was just an additional eighty or so metres of elm that needed to be laid, and this would also be carried out in the Devon style and would take no longer than a week. I sat on the steps of the old wagon as February drew to a close and looked upon the freshly laid hazel atop the bank, silhouetted against the darkening sky as the sun sank behind the moors in the west. As the night chill descended I retreated inside the wagon and lit the woodstove. The wheel of the year was turning and though Christmas seemed just a short time ago, February would soon become March and I would leave the valley here.

8

March

It is said that if March comes in like a lion it will go out like a lamb. I reminded myself of this as a fierce south-westerly charged through the valley. The pitch at Culme was at the southern end of the farm's outbuildings and there was little shelter from the gale, which buffeted the wagon relentlessly. Inside, it felt as though the wind was determined to topple the lorry entirely, rushing beneath her, lifting her weight and dropping her, making the leaf springs groan. I reminded myself that, weighing seven and a half tons, she was unlikely to blow over, but the ferocity with which the storm's gusts caught her bulk made this feel a distinct possibility.

From the window above my desk I could see that Jim had lit a fire in the farmhouse and the thick smoke that rose from the chimney was immediately snatched away by the gale before being dispersed rapidly up the valley and out of sight. I tended to my own fire in the wagon, but just as the flames began to gain momentum they were momentarily extinguished by a rush of wind down the stovepipe, filling the living space with smoke. The fire re-ignited almost immediately and the stovepipe began drawing the smoke once more upwards and out into the valley, but I opened the door to clear the atmosphere inside the wagon, and the wind tried to snatch the stable door from my hand as I did so. I dropped the hook into the latch to secure the open door; it held, but grated noisily, as if trying to break free from its tether. The wind rushed past the open door, although considering the ferocity of the gale, the draught inside was surprisingly light, sufficient to cause a candle flame to dance but not to extinguish it altogether.

The wagon is a welcoming and comfortable home. However, I still find a full day within its confines testing and so despite the gale and sporadic downpours which accompanied it, I decided to make my way to the elm hedgerow below the wood. Elm is common in West Country hedgerows, though the mature specimens that once marked this landscape have been decimated by the arrival of Dutch elm disease on these islands over fifty years ago. In a hedgerow, elm will continue to thrive as long as the hedge is maintained regularly. It is when the maintenance regime is abandoned and the elm hedgerow, like any other, attempts to revert to a line of trees, that it becomes vulnerable to this disease. Once it reaches a height of around twenty feet, the elm will ultimately fall prey to a beetle that will burrow beneath the tree's cork-like bark carrying an aggressive fungus that the elm will react to by sealing off newly infected areas in an effort to prevent spreading. Unfortunately for the elm, this response will ultimately block the tree's ability to effectively draw up water and nutrients and the tree will starve and die.

Since the onset of Dutch elm disease, dead elms have become a common sight across the countryside, and it is only a rare few trees that go on to make thirty feet in height. Maintained as a hedgerow, however, at six or eight feet, the elm remains healthy, and its habit of sending up suckering growth from its roots, in much the same way that the blackthorn does, accounts for many miles of disease-free hedgerow thriving in the south-west. To promote longevity, an elm hedgerow should be maintained as any other — that is, trimmed annually and laid every fifteen or twenty years. Otherwise, the continual cycle of young trees dying from the blight and new, suckering growth emerging from the roots, will lead to thickets of sickly and declining stems rising from younger, healthier-looking growth that, unfortunately, if left to grow on, will inevitably meet the same end.

Elm is one of the easier species for the hedger to work with. Of the different skills involved in the successful laying of a hedge it is the pleaching cut that is the most critical, and learning how each species responds to this process comes through much trial and error. The maples are brittle and prone to crack or snap if the cutting process is rushed; the thorns are easier, though they too will become brittle with age, and so care must be taken to ensure an adequate tongue of wood remains to sustain the laid stem; hazel and dogwood pleach easily and are forgiving. But it is the elm that is the supplest of all the hedging plants. The stem of the elm is leathery once cut, and the tongue of wood retained when pleaching will fold easily and neatly away from the stump. Regrowth from the elm stump will then be vigorous in the spring and the suckering growth will increase the hedge's density still further.

By the late afternoon the storm had eased and I was able to busy myself doing a little preparation on the elm hedgerow, removing the elder and bramble that had found a foothold there. The stream that ran through the valley had become a torrent after the day's rain and it thundered loudly as it tumbled through an assortment of moss-covered boulders, worn smooth by countless centuries of storm water. The pool that lay below the wood churned as the stream deposited the run-off into its depths, and the swirling current collected the dead leaves that had blown on to its surface, sweeping them into a wide spiral before they eventually collected on the pool's quieter western bank, sheltered by a tangle of willow scrub. As I watched the water, I felt sure it must support a population of brown trout, likely small fingerlings that could survive on the nymphs that lived in the pool's gravelly bed, and by rising to snatch insects blown on to the surface from the pasture and woodland.

In the days following the storm I made steady progress and laid the elm hedgerow in good time. The weather gradually improved; though it still came in from the west and brought rain at times, it wasn't the persistent, soaking rain that makes this type of work unpleasant but rather the occasional precipitation that was barely an inconvenience. I watched the weather as it tracked swiftly across the moors; they could be bathed in sunlight one moment and in the next, barely visible through cloud and rain. Winter was on the retreat but was reluctant to leave the field, and spring's victory, although certain, would be hard won.

I had by now been working in the hedgerows for the best part of seven months. As I made the final adjustments to the elm, cutting the stubs from the hazel crooks I had used to pin it to the bank and trimming back any uncooperative stems that refused to stay in line, I felt both relief and satisfaction at the season's completion. The hot afternoons at Eweleaze spent sheltering from the September sun beneath the boughs of oak, with the dark days of winter before me, seemed only weeks ago and yet the greater part of a year had already passed. In less than six months I would once again be amongst the blackthorn at

Eweleaze and the thorn laid this season would be rampant. Even now, thousands of new young stems would be emerging from pleachers, stumps and roots, reaching upwards to enclose the pastures securely, as well as providing shelter for the farmland wildlife that was starting to recover on the holding. The boundary would already be wakening and would soon blossom; the hedger's work does not greatly affect fruiting or flowering, and the refurbished hedgerow would provide plentiful sustenance for the farm's early pollinators, with the newly emerging bumblebees amongst the first to inspect its wares.

The hawthorn at Windwhistle would soon begin to rise under the oaks, and both coppiced stumps and laid stems would be reinvigorated. This plant, most favoured for the enclosure of our lowlands, is also referred to as quick thorn, and within five years the Windwhistle hawthorns should make six feet; several more years and growth would likely be sufficient to craft a hedgerow around a line of stakes. This would make the boundary once again stock proof and complete the restoration. The cottage hedgerows laid at Christmas in the Midland Style, staked and bound with coppiced hazel from the Hampshire lowlands, had caused much interest in the villages. There are few cutters around these districts who work in that style, and though it is not indigenous to the south-western counties the structure's attractiveness has nationwide appeal. The well-laid Midland draws attention to hedgerows and their numerous benefits, as well as showcasing the hedgelayer's craft, and this raised awareness benefits our countryside greatly, regardless of style.

The whips on the hill at Frogmore would have settled by now, and although the driving rain was uncomfortable to work in, it would have benefited the saplings greatly. The fledgling hedgerow would do little in its first few years, but once its virile young roots gained depth, the plants would gain increasing headway and re-establish a corridor for wildlife across the hill's ancient pasture, and so continue the legacy of Frogmore Farm, nestled there on the exposed Dorset coast. My work here on the hazel and elm is finished, the banks on which they rest are ablaze with wild primrose and gorse flowers and will soon be joined by foxglove, red campion and bluebells. The life cycle of these hedgerows now begins afresh, ensuring their place and purpose in this valley remains, unaltered and preserved.

The wagon pulled slowly up the stone drive that led away from Culme, the narrow grass verges on either side by now hosting clusters of daffodils in full bloom. Each flower head resembled the spring sun which now climbed in the east and dazzled me momentarily as I eased the lorry out on to the narrow lane at the head of the valley. My season had finished in good time and I still had several weeks before my first job of the summer began, so had arranged to spend the latter part of the month with a friend who lived and worked from a holding in the north of the county. Jack Bunce ran a modest charcoal and firewood business from his sixteen acres on a North Devon hillside near the market town of South Molton. The holding was divided into two parts with roughly half of it pasture on which he fattened lambs and the other half a small woodland called Whippenscott Copse.

Jack had invited me to stay some weeks back. I had received an unexpected call from him during one of the long evenings at the wood yard; we spoke infrequently, but from time to time he would check in, asking after my well-being, and where I was these days. Jack also laid several lengths of hedge over the winter, and we had discussed work, talking in depth of firewood and hedging, pricing contracts and the relentless wet weather, before he had then enquired of my plans for the spring. With the arrival of March and warmer weather, the trade in firewood eases off as most customers have a sufficient supply to see them through the remainder of the cold season, and so at this time of year Jack's thoughts turn to charcoal. Next month would see the arrival of Easter, and if the holiday weekend coincided with fair weather there would be a healthy demand for BBQ fuel.

Charcoal burning is an ancient craft that has been practised in Britain for millennia. It was widely used in the smelting of iron but with the increased use of stone coal

MARCH

from the eighteenth century onwards the charcoal industry largely died out. Before this, charcoal burners lived an itinerant existence, travelling between the woods and copses where they sourced the wood to feed their kilns, sheltering in huts or tents not dissimilar to the Gypsies who also frequented such places. The process of 'burning' charcoal, as the process of converting wood into charcoal is known, has changed little since this time. The main difference is the introduction of the mobile steel kiln, an eight-foot-diameter ring of steel standing at roughly four feet in height and with a heavy steel lid. The primitive kilns made of earth, grass and moss that were used by the woodsmen of the past have now been largely forgotten, though traces of blackened earth in woodland clearings, now largely unnoticed, stand as reminders of the ancient ritual. These days, charcoal is produced almost exclusively for use as a hot burning fuel for cooking on, and though this market is primarily made up of imported charcoal, a small handful of traditional woodsmen still practise this woodland alchemy in little-known corners of our countryside.

Heading northward, I crossed the River Exe and left the city of Exeter behind me before passing Cullompton and taking the road west, past the market town of Tiverton. Then I crossed the Exe once more as it ran parallel to the

Grand Western Canal. The river here is compressed from a calm, wide pool into a narrow channel, causing the water to be thrown violently between the banks, foaming angrily as it is channelled between the boulders that hinder its progress south to the estuary. The Atlantic salmon still navigates this river, returning to the clear, shallow spawning beds on Exmoor from the cold salty ocean near Greenland where he may have spent several years, adapting continually to the changing environment. He feeds voraciously while at sea and may weigh upwards of twenty pounds on his return, a muscular bar of silver. When he finds himself once more in the river the salmon will cease feeding, his thoughts now focused entirely on reproduction and the continuation of the species. He swims, single-minded, travelling while the river is in spate and the increased volume of water makes his journey less arduous. His silver flanks fade all the while he is in the fresh water, and will eventually dull to the greens and browns of the landscape through which the river flows. At last, he reaches his destination on the uplands. Here, exhausted from the energy he has expelled on his great journey, he at last spawns in the gravel beds near the river's source high on the moor. His work complete, he dies, and his body is given back to the river from which it was first conceived.

MARCH

I left the main road and entered once more into the narrow Devon lanes. I passed through the hamlet of Rose Ash and looking north over the high banks could see Exmoor, which dominated the surrounding country here. The hedgerows that lined the lane throughout the valley on the approach to Whippenscott consisted largely of blackthorn, and already the delicate blossom was turning the hedgerows white, though they were not yet in full bloom. The blossom would peak at the beginning of April and be finished within a month, but even now, before the buds were fully open, the density of flowers occupying each stem made the hedge look as if a light snow coated her form. For country people this is known as the 'blackthorn winter', and refers to the fact that though the appearance of the blossom in the hedgerow is the latest sign to herald the changing season, it will likely still coincide with periods of cold and unsettled weather. On such days, when the blossom is thick amongst the blackthorn's spines and contrasts sharply against a cold, leaden sky, when a biting wind blows in from the north or east and the rooks croak from the still-dormant crowns of oak and ash, the blossom does indeed seem to cloak the hedgerows like winter snowfall.

I approached Jack's house along the newly stoned drive and passed a hand-painted sign informing couriers to 'leave all parcels in the green bin adjacent the old wooden gate'. Jack owned several dogs, one being a smooth-coated Jack Russell terrier who, although now blind, had lost none of his youthful enthusiasm and would bravely bound into the driveway at the sound of any approaching vehicle. This had nearly been his end more than once, and though regular visitors were aware of old Max and knew to avoid him, couriers rushing to make good time had avoided him thus far through sheer luck alone. I pulled up just beyond the house. A steeply pitched black corrugated-tin roof topped the weatherboarded walls of the house, and a set of red deer antlers hung proudly above the entrance. The antlers were spectacular, evenly proportioned and carrying twelve points between them – referred to in these parts as a 'Royal'. They had belonged to an old stag from one of the herds that frequented the woods and pastures here.

The dogs were already barking as I approached the door and as Jack stepped out to greet me he was overtaken by Max the blind terrier, Paddy the Parson Jack Russell and Jess the black Labrador. I was particularly fond of Paddy, who had spent many an hour sprawled in front of the woodstove in the wagon on previous visits. 'All right, bey,' Jack said enthusiastically in his North Devon tones. He grasped my hand in his, which carried the black stains of charcoal, as he had already emptied the first kilnful of the season and bagged it ready for delivery. Once inside, the kettle was put on to boil and Paddy leaped into my lap as soon as I was seated on the old sofa and demanded attention, burrowing his nose into the inside of my elbow. 'Thek old dog is pleased to see ee back,' said Jack, passing me a hot mug of tea. It had been some time since I had received such affection, and I was not used to it.

We began discussing the work to be done. I was to help Jack with two or three burns over my stay, and although the work would not be paid, I would have all the food and drink I needed – and regardless, I enjoyed the company here very much, so there was no need to sully the arrangement with money. Jack was fond of a drink, and charcoal deliveries to the surrounding hamlets and villages would likely include a pint or two in some out-of-the-way inn where, as a well-known character in this district and highly regarded, he was always welcomed as a local.

After tea we made our way outside and Jack directed me to a patch of hard standing that lay between an old oak on the edge of the copse and the charcoal kiln that stood twenty feet above the pitch and was shielded by a makeshift windbreak constructed from old tin sheets that were propped untidily and supported by hazel stakes. Between the pitch and the oak stood a rough outhouse constructed from old gate posts, which had running water and a flushing lavatory, meaning that I would be self-contained for the duration of my stay. The water supply for both the lavatory and the house was piped from a spring which rose on the wooded hillside in the copse and as the pipe exited the wood it passed through a series of filters before finally entering the house. Up until this point the flow was maintained by gravity, but upon entering the house it needed to be pumped to the appliances; the electricity to power the pump came largely from solar panels that Jack had mounted on a barn roof adjacent to the house, though sensibly he had backed these up with a diesel generator that kicked in when the panels struggled during the darkest winter months.

At the point where the water met the filters on the edge of the copse, it was possible to divert the flow into an old enamel bath that Jack had placed here. Though the vessel was pink and clashed horribly with the earthy tones of its pasture and woodland surroundings, it was nonetheless a wonderful place to enjoy a hot soak in the open air. It was even possible to warm the bath by lighting a small charcoal fire underneath. At this time of the year when Jack was burning around two kilns of charcoal a week, his wife preferred him to bathe here, at least initially, to remove the majority of the charcoal dust from his body so as not to soil the family bathroom. By the warmer summer months, a dip in the river at the bottom of the valley was also a viable option.

It didn't take long to get the wagon in order, but by the time I had lit a fire outside, the sunlight was beginning to retreat and the evening birdsong from the copse rang out through the valley. Jack wandered down from the house with the three dogs in tow, carrying two flagons of cider that I recognised as Inch's Farmhouse, a strong, sweet Devonshire brew that I had enjoyed here on previous firelit summer evenings. Sitting around the fire on the edge of the wood, below the charcoal kiln, we talked long into the night, drinking and laughing as we swapped tales of the places and people we had encountered since I had last stopped here nearly twelve months before. Out there in the moonlight, beneath the shadow of the oak, winter was put to bed. Though there were no doubt weeks of inclement weather still to come, now that the hedging season was at an end and the first charcoal burns had taken place, the mood had changed. Thoughts turned to the summer: the grass in the pasture was growing again, the spring equinox had passed and the days were now longer than the nights, the evening fire could be lit outside. The hedgerows, the blackthorn now blossoming and the bright young hawthorn leaves unfurling in the early spring sunshine, were once again the domain of the birds.

I awoke to a cacophony of birdsong and noticed almost immediately the distinctive call of the chiffchaff. The little warbler called its name repetitively from within the woodland, where it was newly arrived, having overwintered in warmer climes south of the Sahara. Heading outside, a brief inspection of the previous night's fire revealed that there was still a good deal of heat in the ash and I was able to rekindle the flames with little effort. Soon I was enjoying a cup of tea while being serenaded by Whippenscott's spring chorus. Shortly, Jack approached with the dogs following him. Paddy ran forward when he caught sight of me sitting by the fire, bounding toward me with great excitement and forcing me to hold my cup of hot tea clear to prevent him from knocking it out of my hand; this however left me somewhat exposed and allowed him to jab his tongue about my face. I couldn't help but laugh. I have nearly always had dogs, and it is only these last few years that I have been without one. For now, I am happy enough to wait until the right dog makes itself known. I would happily let Paddy accompany me on my travels, but his place is here on this hillside, with Jack and his family.

After enquiring about my night Jack produced six eggs from his pockets, still warm, that he had collected from the chicken coop that sat a stone's throw from the house. I ducked inside the wagon to cut and butter bread, and passed Jack a heavy cast-iron skillet that was already primed with a good knob of butter to fry the eggs over the fire. Paddy chose to join me inside, knowing I was a softer touch than his master and that his chances of a morsel were improved greatly by shadowing me. Indeed, after buttering four substantial slices I threw him a crust with a generous portion of butter, which he greedily devoured beneath the armchair, out of sight of his companions who were still fixated on their master. Paddy was a clever dog.

MARCH

Over breakfast Jack and I made a plan for the day. The forecast was clear and the charcoal burner must take advantage of dry weather early in the season as he looks to replenish his stock. Jack had several outlets that had already pre-ordered in anticipation of the increased custom that the approaching holiday would bring; a great many businesses here in the West Country rely on this period after the quiet weeks of winter, and alongside the regular garages and village stores, orders had also been placed by campsites, farm shops and holiday cottages. Several cords of wood lay as yet unprocessed next to the steel drum of the kiln. The cords consisted largely of beech, thorn, oak, willow and ash, much of which had been sourced from hedging work that Jack had undertaken during the winter before hauling it back here to the hillside. Jack is often able to negotiate a deal with the landowner for a cheaper rate on any burnable wood arising from the restoration of their hedgerows. There is little financial profit in either charcoal burning or hedge-laying, despite the labour intensive nature of the work. The reward instead lies in the satisfaction and benefits of physical labour in the open air, as well as giving back so much to the rural environment. Although this work is rarely easy and undoubtedly leaves its mark on the men and women who make their living in this way, it is in my opinion largely beneficial to both body and mind, being closer to the conditions in which we have evolved to both endure and thrive.

It was agreed that I would cut the wood and Jack would load the kiln. Loading correctly is essential for a successful burn, and each charcoal burner has a preferred method of stacking his kiln to produce the best results. Firstly, a good bed for the fire must be made in the centre of the kiln; once complete, this is referred to as the 'charge'. Pieces of wood from previous burns that have been partly carbonised though not yet wholly converted are used for the charge. These pieces are known as 'brown ends', and although they will be graded out when bagging the final product, they still make an excellent hot-burning fuel that lights easily. Once the kiln is fully loaded the charge will be lit, igniting the wood stacked within and setting the initial burn in motion. Four portals situated around the circumference of the kiln serve both to draw air and to provide access to the fuel at the heart of the kiln, and will later hold the four chimneys once the kiln is sealed. Logs are positioned to form a clear channel from each portal directly to the charge, and when the time comes to light it, an oil-soaked rag tied to a length of hazel can be placed easily down each of these channels to ignite the brown ends within. The hope is that the fire catches and burns evenly, distributing the heat throughout the kiln. If the fire burns unevenly only part of the wood will be carbonised and this will result in a much-reduced haul.

I set about processing the wood, cutting the logs to around a foot or eighteen inches in length before passing them to

MARCH

Jack, who having first carefully constructed the channels now proceeded to stack the wood in a spiralling pattern within the chamber. The larger logs and denser hardwood he put at the centre, while the smaller, softer willow he stacked closer to the kiln's edge. It took most of the morning to cut and load the drum, which held a considerable amount of wood, and it was gone twelve by the time we had finished. The stacked wood now rose just proud of the kiln's rim and the heavy, conical lid which would eventually seal it remained propped open by two lengths of oak and resembled a mouth agape. The whole construct now sat dormant, awaiting only the lighting of the charge to bring it to life.

The fire at the wagon that had provided us with a steady supply of tea for the morning had by now burned to ash, and we decided to head back to the house for lunch. Jack had built this house over the course of a summer, having previously lived with his wife and two children in a static caravan that had been dug into the hillside. The house was constructed entirely of wood and the considerable floor space was polished concrete on which sat a large rug that made the open-plan living area feel homely and comfortable. From the cathedral ceiling hung a chandelier made of deer antler from the roe and red deer that frequented the land here, and which Jack hunted for the table. In the kitchen space at the rear of the living area, a huge farmhouse table spanned almost the width of the room. This downstairs also housed a bedroom and bathroom whilst the children had a bedroom each upstairs. It was a beautiful home, tied to the land and woods here. Jack was a born and bred Devon boy and had a long heritage of farming stretching back many generations; his great-grandfather had farmed sheep on this very hill and his mother still lived in the family farmhouse in an adjacent valley. He himself had returned to the land after several years serving in the forces and had found the peace and solitude he craved here on the hillside.

After a good lunch, time had been set aside for deliveries before the kiln would be lit in the late afternoon. The burns in the early season when the nights are still chilled can take anywhere upwards of twelve hours and the kiln would be supervised throughout the night and closed down the following morning. Twenty-four hours after that the charcoal would be ready for grading and bagging.

By five o'clock in the evening we had returned to the kiln. The sun was low in the sky and illuminated the proliferating white blooms of blackthorn blossom in the valley below. There is a sense of occasion in lighting the charcoal kiln, perhaps because it is such a timeless practice and often carried out in ancient woodland overseen only by the animals and birds and the trees themselves, unmoving and time worn, ever watchful of the woodsman's habits. Due to the hour, tea had given way to cider and we both raised a glass to a successful burn. Jack then took four hazel sticks he had recovered from his barn, all relatively straight and cut at about five feet in length, and wrapped some ripped rags around the ends to create torches. I had lit a small fire next to the kiln, and after soaking the rags in old vegetable oil, Jack left the torches in the flames until each had caught. One by one he then proceeded to push the flaming torches into the portals and through the log channels he had created that morning, waiting until the torch came into contact with the brown ends that lay at the heart of the kiln. Once all four flames were in place, Jack stepped back, extinguished the small fire he had used to light the rags, and reached once more for his cider. Now we waited.

Slowly the fire grew. At first just a hazy wisp of smoke escaped the spiral of wood in the drum and the pop and crack of the young fire taking hold in the belly of the kiln was barely audible. Jack kept a close eye on the portals, feeding the hazel sticks ever deeper as they burned and making sure the fire had established in each quarter. The noise inside began to increase, the cracking was now accompanied by a constant hum as the flames within began to gain momentum, the delicate wisp grew thicker and rose more urgently as heat was cast out from the drum. Within ten minutes the kiln was erupting with great plumes of smoke and the sound had increased in intensity, the flames roaring unrestricted and the cracking within the drum sounding like bones snapping in its ferocity. Jack turned to me. 'Us better get thek lid on, bey.'

The fire had reached sufficient strength and now the charcoal burner would exert his authority over this ancient ritual, calming the flames, quelling their appetite and bringing it to heel.

We positioned ourselves on either side of the kiln, our view of each other hindered somewhat by the thick smoke that swirled all around us. On Jack's word we each drew our wooden prop from the lid which fell into place with a heavy thud. The fire gasped in shock and leaped momentarily from the portals at the kiln's base before retreating once more to its lair within the chamber. Next, the chimneys were inserted into the portals in each quadrant and all began to smoke steadily, a good sign that the fire inside was well established throughout. Now the last gaps were sealed. We shovelled earth on to the lid and patted it firmly around the rim to ensure an airtight seal, any stray tendril of rising smoke alerting us to where more earth was needed. We banked the earth up around the kiln's base and it was possible to see the loose, dry earth visibly moving as the fire within attempted to suck more air through the diminishing gaps. Eventually, the only air supply available to it was drawn through a small crack on each quarter under each chimney, just enough to prevent the fire inside from suffocating and to sustain it over the next twelve hours or so.

MARCH

The light faded, the kiln's white smoke drifted silently away into the copse, and all the while the wood inside the locked chamber cooked. Within the steel drum, the intense heat was transforming the wood, driving from it the different elements from which it was made and returning them to the air, until just one element remained: pure carbon. Outside, Jack, well practised in this alchemy, carefully tended the embers that fuelled this ritual. The night was calm and still, the call of the tawny owl the only discernible sound in the valley as we kept watch. The kiln, too, was quiet now, though the heat that radiated from its steel walls told of the frantic activity contained within. Jack continued to tease the fire, keeping the embers starved of the air which would see them flare up and consume the wood inside, allowing them to taste but not devour it.

As the ritual continued, the wood inside wept tar, which threatened to block the portals and stifle the fire altogether, and so at intervals, the chimneys had to be rotated and the portals checked to ensure the burn was even and thorough. We decided to take this task in turns, checking the ports every four hours. It was now eight in the evening, Jack would check them at midnight, I would do the same at four in the morning, and we would reconvene at eight for breakfast by my fire. I sat with Jack for another hour before turning in. The sky was clear and unpolluted by the lights of town or city, the stars were bright and sharp, and for a moment, sitting there around the kiln, it was as another time. I have rarely felt as connected to the landscape as I did that night burning charcoal on the hill.

I rose from my bed at four. The stars were now largely obscured by cloud cover and a slight breeze had moved in, preventing a frost from settling which had felt likely the previous evening. The kiln was smoking evenly from the four chimneys, the fumes rising up the hill in the early-morning air, their drift signifying a change in the weather; the wind had swung round once more to the west. I put on a thick pair of welder's gloves which had been left by Jack hanging over the tin sheets that sheltered the kiln from the westerly. I then proceeded to rotate the chimneys, checking the portals for tar and residue and making sure the fire remained sufficiently contained. With the time only just gone four, I went straight back to bed as soon as I had completed my task; though the air inside the wagon was chilled, my bed was still warm, and I fell back to sleep quickly.

When I awoke shortly after seven it was already light, though the sun was absent and the sky a pale shade of grey, and the birds were in full voice. The kiln was still burning evenly as I collected brown ends to make a fire outside the wagon, and though I noticed the smoke that rose from the chimneys was now thinner, the ritual was almost complete.

Jack soon joined me, Paddy running ahead to greet me, enthusiastic as ever. We drank tea by the fire for perhaps an hour while the smoke from the chimneys gradually became more opaque, tinted with a slight shade of blue that signified the burn was complete and the fuel now burning within the belly of the kiln was not wood but charcoal. It was time to extinguish the fire.

We removed the chimneys and sealed the portals by shovelling earth into the last of the gaps, cutting off the trickle of air that had sustained the burn until this point. The kiln radiated intense heat from its steel frame. A cup of water thrown on to the lid of the drum at this stage would evaporate almost instantly, and though the fire inside was now dying it would not be cool for another twenty-four hours. It would be another day before the lid could be safely lifted without fear of re-igniting the flames; only then could we inspect the hoard. The burn cannot be considered a success until the quality of the charcoal has been seen, and for now the kiln would keep us waiting.

The next day Jack joined me for breakfast. The weather, though cloudy, was milder than recent days and it was comfortable outside. More importantly, it was dry, and with the kiln now cooled the charcoal within could be graded and bagged, before the drum was reloaded and the ritual repeated. Jack cleared the earthen seal from the lid, then we each took a side and dragged the heavy, conical cap back and clear of the kiln. We propped it against an old willow that grew from the bank on the edge of the copse, which was by now adorned with the distinct downy catkins that had given rise to its common name, the pussy willow. With the lid stowed safely we peered in at the kiln's contents. Each piece of wood was preserved exactly as it had been when the kiln was loaded, every knot and bud still clearly visible, but it was now transformed. Picking a piece up, it was as light and delicate as fine bone china in my hand, and when it caught the light it was pearlescent like the wing of a starling. Pure carbon.

At a time when we seek to reduce this element in our atmosphere, its purposeful manufacture through an ancient ritual may seem outdated, perhaps even destructive. However, like the hedger, the charcoal burner is bound deeply to his environment, a cog in a wheel that has turned for millennia, and his habits seek to preserve a sustainable way of living and working. The woodland and trees he relies on for his livelihood are both cherished and maintained in a careful balance. The charcoal burner harvests wood from derelict hedgerows and coppices, thinning the woodland and allowing light once more to reach the shaded floor, which encourages the growth of dormant saplings and seed, prompting the flowers to bloom, and even the cut stumps will soon send new shoots upwards. The woodland is greatly invigorated by the activity of the charcoal burner. The tourists cooking their barbecues whilst holidaying in the West Country are for the most part unaware that the fuel

MARCH

they are using to prepare a meal, if sourced thoughtfully from a local manufacturer, is of such benefit to the woodlands here.

The burn had been a success, and the kiln bore a good load of hot burning charcoal that would now be graded before being bagged and delivered. The grading of the charcoal is an essential part of the process. A mesh frame is propped against the kiln and the charcoal is passed over it; the 'fines', pieces too small to bag and sell, will fall through the mesh and will eventually be dug into the vegetable garden, locking the carbon away and improving the soil. The more sizeable lumps that remain will then be sealed in paper bags to be sold as lump wood charcoal. It was my job to get into the kiln and shovel the hoard on to the waiting mesh, with a thick hankie secured around my face to prevent me inhaling the fine carbon dust that flew up. Jack, also wearing a mask, then dragged the char across the mesh by hand, and into a waiting bag which hung pegged open on a purpose-built frame. As he brought the charcoal across the mesh, he scanned each shovelful for brown ends, snapping any larger lumps to be sure they had converted all the way through before placing them into the bags, all the while combing the hoard with his fingers to ensure the fines dropped through the steel mesh.

It was lunchtime by the time the contents were bagged,

and the pair of us were black all over, like coal miners returning to the surface after a long shift down the pit. We shared a simple lunch of eggs, boiling them in an old and soot-blackened saucepan over a fire of brown ends, enjoying them with a sprinkle of salt. The afternoon was spent making deliveries, and at the end of the day we were both still coal blackened. Jack's wife made an exception and allowed him to get cleaned up in the bathroom, after wiping the worst of the charcoal from his face and hands. I made use of the enamel bath on the edge of the wood, lighting a fire underneath to heat the water through. It was peaceful in the valley and I lay back in the hot water and took a long draught of cider. The birdsong rang out from the coppice as afternoon became evening, and I was perfectly content in that moment, there under the willows at Whippenscott.

9

April

It was the turn of the month as I drew the wagon slowly up the stone track toward Marsh Farm. The elm hedgerow that lined each side was just coming into leaf and the blackthorn that had gained a foothold here and there was conspicuous, each stem now heavy with white blossom. Upon entering the yard, Tim and Susanne approached the wagon with warm smiles and greeted me, 'Welcome home, Paul.' It had been busy on the farm in my absence. The sheep had begun lambing in February and home field was now lively with young lambs; unrestrained and full of life they grouped themselves into spirited gangs that romped through the field with a devil-may-care attitude. It was a scene that couldn't fail to lift one's spirits.

After exchanging pleasantries, Tim told me that the ground on which I had stayed over Christmas had now been set aside for the construction of a new barn. I was still welcome to pitch the wagon there, he said, but building work was due to commence imminently and there was a potentially better alternative. Toward the back of the farm, on ground he referred to as 'the Marsh', was an old cider orchard that was accessed by a hard track wide enough for farm machinery, which would be no trouble for the wagon. Despite the area's name, the ground closest to the hedgerow which bordered the track was hard and compact and would be suitable for a pitch, certainly over the spring and summer months. There was some work to complete here, too. The old wooden fence and gate that enclosed the orchard were in a state of disrepair and needed replacing, and Tim, who was now busy with both the farm and the lorries, had little time and would appreciate the help.

And so, shortly after my arrival, I headed down to the Marsh, navigating the winding and gravelled track that led me toward the River Parrett which marked the far boundary of the holding. I followed the route through lush grass fields peppered with sedges and past a line of archaic willow pollards, the slender oval leaves just beginning to emerge alongside the trees' yellow catkins. The willow thrives in Somerset's reclaimed pastures, and the ancient and fissured trunks that lined the track may have stood on these marshes for centuries. Each gave rise to a multitude of straight stems that were themselves now aged and heavy. These stems would once have been harvested when still young and supple to make the hurdles used to enclose the sheep that were reared

APRIL

here, but these veterans had not been pollarded for many decades.

Eventually the orchard came into view. The apple trees were timeworn and of great character but were still bare, their blossom remained several weeks away and would coincide with the return of the swallows from Africa. These agile birds would currently be preparing for their great migration northward and some early pioneers might already be on route. Presently, I came to a place where several alder stems rose from the hedgerow, a little more than fifty yards from the orchard's failing gate, which hung unceremoniously from a leaning and rotted post and was only prevented from falling completely by several loops of faded orange nylon bailer twine which had been tied to an ash tree growing from the hedge. I got out of the wagon and inspected the gateway, making sure the ground there was firm enough to turn the vehicle around before committing to the manoeuvre. Satisfied with the ground conditions, I decided I would make my pitch there, under the alders and with the stable door facing east, towards the gate and orchard beyond and sheltered from the south-west wind. To the south beyond the hedge lay a field of corn, and though barely a couple of inches of green showed from the ground now, it would soon gain momentum in the lengthening days and warming soil. Beyond the cornfield was Cannington, the church steeple rising clearly from the cluster of cottages and houses that were just visible.

Dominating the background lay the Quantock Hills, which climb up from the Taunton Vale and continue westward until they meet the muddy waters of the Bristol Channel, west of the Marsh, at West Quantoxhead. It was quiet here, away from the farm buildings, and although the sound of the village and the roads that led to the town of Bridgwater several miles to the east were still audible, the overriding feeling I had there was of being cut adrift; there was solitude to be found here on the Marsh.

A few days later I made the journey to Dorset to collect my pickup. George was processing firewood in the yard when I arrived and the afternoon slipped by as we talked. Those of us who work in the woods and hedgerows spend many hours alone, and when a meeting of any kind is arranged it will rarely leave time in the day for much else. Although most woodsmen I've met and worked alongside are fond of their own company, they are all capable of talking for many hours, and the days and weeks spent pondering any number of subjects in solitude will undergo thorough analysis and discussion when we find ourselves in company. Though small talk is largely dispensed with, farmers being especially known for their straightforwardness, when time allows and the company is amicable, many hours can easily be lost. In my experience these hours are well spent, and the knowledge I have gleaned of the woods and hedgerows has come less from books than from experience and from listening carefully to the men who came before me. The way of the woods and

APRIL

hedgerows is very much an oral culture, and in recent years it has been shared among a dwindling community.

The hedging season was now behind me and for the next five months I would be staying here by the orchard on the Marsh, helping out on the farm between my own work and the occasional foray abroad. I would likely travel to Devon and Dorset later, in the summer, once the hay was in and stacked in the Dutch barn. Each year in the spring I undertake several contracts installing wattle fencing, weaving hazel rods between upright stakes to create a continuous barrier that flows with the contours of the ground on which it sits. The style is not dissimilar to the hazel sheep hurdles that were produced en masse to create pens and races for the great numbers of sheep that were kept in the heyday of the wool trade, and which saw thousands of acres of hazel woodland worked and maintained to produce the long, straight rods required. Like wattle fencing, these were made by weaving hazel rods between uprights to produce a sturdy barrier measuring roughly six feet by six feet. Hazel hurdles are relatively lightweight and manoeuvrable and could be easily manhandled by the shepherds who tended the vast flocks across the south of England. Eventually, as the wool trade diminished and the mass production of metal alternatives made hazel less attractive, many hazel coppices fell into decline and much of what was not grubbed out was abandoned, left to form dark, forgotten thickets that served little purpose beyond sheltering deer. Thankfully, not all was lost. A managed hazel coppice, apart from being a rare and precious habitat frequented by some of our most elusive mammals and invertebrates, is also a most suitable environment in which to raise pheasants, one of the country's most popular game birds. And so a fraction of this woodland habitat was preserved, and is still worked today.

I am a regular visitor to the hazel coppice in the months of spring, buying in rods by the hundred, cut from traditionally managed woodland, some of which has been worked continuously for hundreds of years. The wattle work I produce uses the top weave familiar from laying a Midland hedge, three rods being worked at any one time and woven between upright stakes, gripping them tightly and creating a barrier of not inconsiderable strength. This woven 'rope' worked in alternating directions builds an attractive herringbone pattern and sits well in the cottage or country garden.

The coppice I frequent is a two-hour drive from Marsh Farm, heading eastward, skirting the Dorset border and tracking across the county of Wiltshire to Hampshire beyond. After crossing the Salisbury Plains and passing Stonehenge, which rises to dominate the surrounding grassland and ancient barrows for which the plains are famed, I leave the main road. From here I pass through an arable

APRIL

landscape. Wheat, barley and oilseed rape are cultivated here on the white chalk earth, and early golden flowers are already peppering the vast fields of rape that within a month will turn broad tracts of this landscape a buttery yellow. Eventually I descend into the market town of Stockbridge, nestled in the Test Valley. The main street passes directly over the Test and its clear, chalk-filtered waters. The river rises from the hamlet of Ashe in the north of the county and supports a sizeable population of our native brown trout. Its reputation as a premier location for fly fishing now stretches far beyond our shores, and the river is frequented by country gentlemen drawn to the quality of its trout fishing.

Beyond Stockbridge I follow the road eastward and into the ancient oak woodlands of Hampshire, where the hazel is still managed for the rural craftsman, for hedger and hurdle maker, thatcher and basket maker. Leaving the metalled road, I climb a gravel track into the woodland, where ash and oak trees rise from the carefully managed understorey, casting just enough shade to draw the hazel rods and poles straight up without hindering their growth through lack of sunlight. Following the track, I startle a group of fallow deer who leap across my path. With no natural predators, the increasing deer population is a concern to the woodsman, as their habit of grazing on the delicate new shoots that grow from the cut coppice stools can, in extreme cases, see acres of woodland rendered useless. On occasion, the woodsman culls the deer, the burden of thinning their numbers to a sustainable level coming not solely from a desire for sport, but rather to maintain a balance upset in earlier times, when the destruction of large predators was seen as beneficial to our advancement as a species.

The countryside, its pastures and woodland, have been shaped and managed by humans since our ancestors abandoned their nomadic existence as hunter-gatherers amongst the wildwood and grasslands. As people settled and agriculture became the primary tool for sustaining the population, the landscape was changed. Man's ability to manipulate nature for his own gains did not come without consequence, and these changes have affected ecosystems and food chains, at times upsetting a balance that took millennia to evolve. If in the past these changes were considered acceptable, even necessary to our own progression, more recent years have seen a shift towards a more enlightened train of thought whereby we manage the countryside not solely for our own gain but also for the preservation of habitat and species that have declined since we appointed ourselves custodians of the land and its other inhabitants. In the case of deer culling, the aim is to maintain a healthy and sustainable population; to achieve this, the woodsman targets the old and injured, as bear and wolf would have done in our past before humans pushed these species to extinction on these islands. He now fills that ancient role of hunter and it is in this way that nature has evolved; it is likewise in this way that we preserve

APRIL

what remains of the land's oldest and most revered landscapes, habitats and species.

As I progressed through the wood, the understorey changed. Earnest Benham, who worked these woods, cut and processed roughly an acre each year and driving along the woodland track the differing ages of the hazel were clearly apparent. Upon entering the copse the hazel had been tall, perhaps twelve or fourteen feet in height, and each coppice stool threw up between twenty and forty rods which would soon be ready for harvesting. As I passed through each section or 'coupe', as they're known, the height varied depending on its age and place in the rotational cutting cycle. The density of the stools, however, remained consistent throughout the copse and each was just shy of a couple of paces away from each other. Eventually I reached the latest coupe, where the stools were freshly cut, and just beyond here lay drifts of neatly stacked brushwood which were yet to be processed. All along the track delicate primroses carpeted the woodland floor, their growth so prolific that it was probably possible to cross the entire copse, touching one with every other step. Bluebells, too, grew thickly and wrestled with the primroses for space. Most of the bluebell flowers were still closed, though the density of plants hinted at the spectacular display to come in just a few weeks. Some of the flower heads had already taken on the distinctive violet hue as they began to ripen and several more amongst their number were already nodding. It was impossible to avoid the flowers when piling up the wood, although Earnest had done his best.

Bundles of faggots lay tightly stacked, two and three high, more than a hundred in total. These would be collected by lorry and distributed among a number of environmental organisations, many of whom preferred this organic woodland product to the stone-filled gabion baskets so popular in recent decades for riverbank restoration and habitat creation. There were over one thousand hedging stakes that Earnest had taken the time to cut and stack carefully, all facing in the same direction, the thicker butts at one end, barely an end protruding untidily from the pile. Though the hedging season was over, these stakes would keep well here in the wood, and were in great demand with the hedgelayers whom Earnest supplied, as they were consistently straight and uniform. Rods reaching twelve feet or more in length had been tied in bundles of twenty, graded according to thickness and stacked vertically to extend their longevity. Cut over the winter and early spring when the sap was down, these rods would remain malleable for months and would be suitable for weaving with well into June. The thicker rods would be split into two by the hurdle maker who would then use them for his woven panels while the thinner rods, to be woven 'in the round', I would use for wattle work. Finally, there was a pile of stout sticks that were

APRIL

largely knot free and had been cut from the first length of the cleanest rods. These, Earnest would fashion into spars for the thatcher, processing them here in the wood under a faded green, waxed cotton tarpaulin tied between oaks and propped roughly by forked hazel poles. There was normally a fire burning in this makeshift shelter and fresh tea was always to hand. Alongside a good dinner, it is generally strong tea and perhaps a little baccy that sustains the woodsman in his labours. In fact, just a little way along the track I could see smoke rising from below one of the old oaks, and as I drew closer I saw a kettle nestled amongst the flames of an open fire. Earnie was expecting me.

I hadn't known Earnest for long; it was only eighteen months previously that my continuing search for good-quality hazel had first brought me to this coppice. The standard of the wattle work I produce is a direct reflection of the quality of the rods used, and for this reason many others who create similar structures choose to work in willow. The willow grows quickly, in less than half the time it takes the hazel to put on the same growth, and is often cut on a two-year rotation as opposed to the hazel's seven. Then there is the fact that willow rods are generally gun-barrel straight and perhaps easier to work with. However, a 'withy bed', as a willow plantation is known, will support far fewer species than the ancient woodlands where the hazel is cultivated and though the two plants have much in common, it is our native hazel, Corylus avellana, and the coppice woodland where it is managed that most benefits our countryside and its biodiversity.

I had taken an immediate liking to Earnest. Within minutes of entering into conversation it had become apparent that he had spent decades working the woods. He did not need to tell me this, it was apparent in his accent and his face, which was nut brown and weathered and spoke of a lifetime spent outdoors. His hands, too, were fissured and calloused from countless days spent gripping tools, and they didn't stop moving, shaping his hazel spars effortlessly while Earnest talked to me, gauging my own background before he became less guarded. The way Earnest seemed to mould into that environment like the deer or hare or birds told that he was a woodsman in the truest sense of the word. He belonged here.

Upon my arrival, Earnest put down the chainsaw he had been using. This is now the preferred tool for the coppice woodsman, as it is for the hedger, though axe and billhook are still favoured for dressing out the harvested rods. Removing his helmet, he gave me a broad smile. 'How you bin getting on, Paul?' he asked. 'Oh, struggling on, Earn, bearing up under the strain,' I replied. He laughed as we shook hands firmly. It is customary to at least feign some hardship when meeting a fellow woodsman, especially when you intend on purchasing a considerable amount of his produce.

Earnest made us both a cup of tea, snapping a stick from

APRIL

a nearby hazel and using it to remove the teabags. The spar trade was picking up, he informed me. Recent years had seen a fall in demand as cheap thatching spars imported from Eastern Europe had increased, but with these imports now harder to come by, Earnest was receiving enquiries from thatchers all over the south. 'I've 'ad to turn custom away, Paul, though I hates to do it,' he informed me. 'I would rather say no than let people down.'

Like hedgelaying, thatching is an old country craft that, carried out well, accentuates the beauty of our countryside, and the sight of a well-thatched cottage is one of the first images that springs to mind when picturing a scene of rural England. Thatching was traditionally carried out by rural labourers who thatched hayricks, and when neither harvest nor cultivation demanded their attention, they would use these same skills to restore any failing cottage roofs on the farm or estate. It was a skill that was known to many countrymen and was, perhaps, taken for granted, but as the years moved on and agricultural practices changed, satisfactory workmanship became increasingly hard to come by. The increasing mechanisation of agriculture rendered manpower both expensive and less productive, and a significant cohort of the rural labour force left the land to seek employment elsewhere, resulting in a decline in skilled craftsmanship in the countryside. However, with a quintessentially English country cottage desirable to many seeking a country idyll in our more rural counties, the thatcher's skill remained in demand, and it is one of the few crafts that has been positively impacted by the countryside's modernisation. Unlike many other heritage crafts, there is now a guild that oversees the national standard of thatching, preserving the skills and ensuring that the quality of craftsmanship remains high. A reputable contractor will be accredited by this guild and known as a Master Thatcher.

The thatching industry, therefore, still has need of woodsmen like Earnest, who prepare the hazel spars and 'liggers' needed in the construction of the roof. Earnest made thatching spars by the thousand, splitting hazel wands with a spar hook designed specifically for this purpose. He possessed the gift common to any master of their skill, and that is to make the process look easy. Each spar that Earnest produced was identical in thickness and form, regardless of any irregularities in the wood with which he worked. He also made them at a considerable rate, his hands so accustomed to the process that he barely had to think about the task ahead, and was able to talk easily while he worked, pausing only to sip his tea. As I watched him I was reminded of the village women who sit knitting while chatting amongst themselves, a beautiful scarf or jumper flowing from their hands as they catch up on village affairs, never dropping a stitch, the pace of their task determined only by the excitement of the gossip.

I stayed with Earnest for two hours or so, sitting under the oak and making several cups of tea as we talked. All the while his hands were busy with the spar hook, nimbly

APRIL

splitting the hazel wands before pointing each end with three clean strokes of the tool, its edge as sharp as a new razor and wielded with such skill it was like an extension of his arm. We agreed on a price for each bundle of rods. With the availability of such prime hazel in short supply he was able to ask a premium price for his product, though I did not begrudge paying what he asked, nor did I quibble. I knew that neither of us would get rich from our labour.

Together we loaded the rods on to the pickup; we managed to get seven hundred rods in all tied down securely, though they overhung its dented bodywork both front and back. When I left to make the return journey to Somerset the fully laden pickup looked even more conspicuous than usual and it drew much attention as I navigated through Stockbridge and onward to the A303 beyond. Both the hedger and his work are still relatively unfamiliar to those not working the land in some way, and to many Hampshire residents my truck, engulfed by hazel, must have looked almost like a carnival float.

As I crossed the Marsh, I drove carefully past the old willow pollards where the track was at its most rutted as I didn't want to risk upsetting my cargo, just a few hundred yards from my pitch. A heron suddenly launched himself skyward from the ditch beyond the line of willows, where he had been stalking the frogs that populated the network of ditches criss-crossing the marshes. The heron would likely by now have eggs in its nest, which was located with several others in one of the larger pollards beyond the orchard. I had spotted the heronry earlier in the month, when the nests were clearly visible amongst the willows' juvenile foliage. These handsome birds inevitably give their position away, being a considerable size and easily identified by their croupy calling and awkward manoeuvrings amongst the canopy. The adult birds were now busy throughout the day working the ditches, and their evocative call was still audible after nightfall and would often carry clearly to me as I lay quietly beneath the alders waiting for sleep.

The wattle fencing work kept me busy. Its rustic style suits the cottage gardens of the West Country, whether as a backdrop to a flower bed, a woven border around a lawn, or a screen to hide unsightly oil tanks or refuse areas, and it is much in demand. In basketry, I believe the weave I build is called a 'three-rod whale', though I have always known it as a Midland hedge bind, as the method uses the same structure put in place when laying a hedge in that style. The uprights are equally spaced at the distance between elbow and fingertips, which is around half a metre. Once these uprights are installed the weave can then be built up in layers, either working alternately in opposing directions to create a herringbone pattern or binding consistently in the same direction which provides a less intricate pattern though is

APRIL

still very attractive. The screen or fence follows the contours of the ground, and in my opinion is particularly attractive when flowing through an undulating garden where the curves and waves accentuate the natural fluidity of the hazel.

I had been asked to weave some of this fencing at a property where I had laid two sections of hedgerow before Christmas. The two blackthorn hedgerows which I had worked on ran at right angles and the customer now wanted to install a gate where they met, giving access to the kitchen garden. The blackthorns I had laid were now past their first blossoming and were already shooting from the cuts made just four months previously, and the woven hazel binding that ran along its length was still clearly visible. I had come up with a structure that I thought would sit well between the hedgerows: a framework of oak enclosing split hazel for the gate, topped with a woven hazel rose arch.

The village where the work was located was not far from my pitch by the orchard, and lay just half an hour or so to the east, beyond the town of Glastonbury and amongst the farms and cider orchards that sit in the shadow of the Mendip Hills. I decided to make both the gate and the arch here on the Marsh, and was confident the completed structures could be easily transported and installed in place within a day. With both items built it would be the digging of holes in which to sink the gate posts that would be the most strenuous and time-consuming task.

I started the week by journeying over the Quantocks to

a sawyer I knew who ran a sawmill that processed oak, cedar and pine, not far from the village of Stogumber in West Somerset. The scene was typical of the small independent sawmill: a dilapidated old timber shed kept the worst of the weather from the person operating the mill, although the deck on which the round timber was placed stretched beyond the building's shelter and into the wood yard beyond. The yard itself was stacked high with the cedar logs that were awaiting processing; the bulk of these would be sawn for exterior cladding. The cedar's natural resilience to rot makes it particularly suitable for this job, and indeed the exterior of my wagon is clad with cedar that has proven to be most durable. Another open-fronted shed held the sawn timber. Oak logs that had been milled into boards and beams would be used for barn conversions, fireplace and window mantels and landscaping projects, and these sat propped on bearers and gathered dust whilst they awaited sale. Cedar cladding and building timber were also stacked in the shed, which I noticed had several swallows' nests on the back wall, nestled snugly against the joists amongst ancient cobwebs that now hung heavy with sawdust. Within weeks, perhaps days, the swallows would again be darting around the buildings, collecting the mud from the edges of the ruts that lined the approach to the wood yard and using it to renovate their nests.

I was here to collect oak for the frame of the gate as well as some oak uprights that I would use for another fence in the same village. Although hazel is cultivated for hurdles and wattle work, it contains none of the tannins to be found in either oak or sweet chestnut and so longevity can be increased by using these hardwoods to support the weave. Even then, by five years of age the weave will likely be brittle and deteriorating, although a coating of linseed oil applied twice a year should extend this lifespan by several seasons. The nature of the product is such that it will deteriorate quicker than imported wood species that have been pressure treated with chemical preservatives, and obviously quicker than plastic borders and fences of which, to my dismay, I have seen many in recent years. Hazel wattle work is however unrivalled in both its aesthetic appeal and its beneficial role in preserving British native woodland habitat, and for this reason it continues to be a popular choice.

The sawmill was run by a Somerset native, a tall and wiry man, perhaps ten years older than me, by the name of Otto. Before taking on the mill he had been a woodcutter, spending many years cutting timber throughout the west, as I had. Woodcutting is physically demanding and though recent years have seen the work increasingly mechanised, there are still many men whose early career was spent amongst the pine and beech, the spruce and ash, felling and processing timber with chainsaw and steel wedges for the pulp and sawmills. I warm to these men; they are a breed who spent their prime grafting for little financial reward, travelling between contracts and living in caravans under the woodland canopy. Their lives in many ways mirrored the itinerant existence of the old charcoal burners, foresters and hedgers, moving with the work and seasons across our countryside. This heavy and demanding labour will eventually take a toll on the body, however, and even if they manage to avoid injury, the employment will have had the best of a man by the time he reaches forty. These men, now advancing in years, often stay within the industry, the wood and sawdust having become an integral part of their make-up, and go on to be employed as machine operators or sawyers, hauliers or timber merchants, retaining a connection to our woodlands.

I talked with Otto for some time, drinking tea from a chipped mug whilst he cut oak fence stakes on his re-saw. He cut several more than I needed at no extra charge to compensate for any that might warp or split before I had set them in the ground and woven the hazel between them, as oak has a tendency to twist and move. The best part of the day was already behind me when I returned to the Marsh, and by the time I had lit a fire outside the wagon and waited for the kettle to boil, the light was beginning to retreat. I sat on the steps, the evening air feeling pleasantly warm, and with only the noise of the lambs on home field and the occasional squawk of a heron, it was peaceful on the Marsh.

APRIL

I awoke to a gentle spring rain that tapped lightly on the window above me. As I gradually came to, I eased myself up on an elbow to survey the day and could see the sun rising to the east. Once I was up and dressed the shower had passed, so I went to light a fire outside. I had plenty of firewood stacked beside the hedge, consisting mainly of alder and ash from the hedgerow, but also some fallen applewood from the orchard and even some of the old softwood fencing that I had dismantled and was yet to replace. I had so far removed some of the ironwork from the old gate posts, pulling the rusting nails from the timber before using it for firewood. The heavy oak sleeper which had formed the gate post was now charred from the previous night's fire, but would no doubt still hold some of the heat in its bed of ash. I rolled it aside and added more kindling, and soon the flames had reignited and the kettle was boiling.

I sat there for some time drinking tea as the sun gradually gained height and lit up the orchard before me, illuminating the first blossoms that were emerging as the month progressed. As I hung the kettle back on the iron above the fire, completely absorbed in the task at hand, I became aware of a distinctive, shrill chatter. Casting an eye upwards I saw a familiar display as a group of swallows raced through the air around me. The unrivalled pace with which they cut through the April sky, and their self-assured manoeuvring, propelled them effortlessly in all directions; they appeared unbound and unhindered by any earthly shackle, as free in life as most beings can only hope to be at the end of life's journey, when worldly burdens become insignificant. The return of the swallow is a joyous moment in the year for country people and townsfolk alike. All who have been working the land over the long winter, as well as many whose lives are perhaps more detached from the ebb and flow of nature's rhythms, cannot fail to be uplifted by this most recognisable sign of the progressing year.

Preparing the framework for the gate was a relatively straightforward task and I was done in good time. I now had a rectangular surround, three feet wide and four feet high, with two black steel hinges screwed securely at the left side of it and a latch at the right. Dividing the space within the frame were two upright hazel rods, which I had whittled down with a pocket knife at each end to make sure they fitted snugly inside holes I had drilled into the oak frame; these would provide the support for the weave. I now set about splitting the rods that would fill the space. This is a task that takes some practice, though it is achievable to even a novice if the process isn't rushed. I am a hedgelayer and though I work regularly with hazel, I am visibly off the pace when compared to either hurdler or spar maker. I do, however, take great satisfaction in the process and was quite content to sit there on the Marsh, next to the fire, as I

APRIL

carefully split the lengths in half before weaving them between the uprights within the gate. By mid-afternoon the job was completed, each of the hazel lathes held tightly between the hazel dowels and filling the body of the gate nicely. One side now showed the attractive, mottled silvery bark, whilst the other displayed the creamy white of the hazel's heart, all enclosed within its oak frame. It was an attractive structure, and I was pleased with my work. The oak and hazel are often companions in the woodland and complimented each other perfectly in this rustic assembly.

As evening fell I rolled the charred oak gate post back across the ash of the fire before adding the hazel splinters and chips scattered about me from the day's work. I stirred the hot embers to expose them, before blowing on them gently at which they ignited and I was able to add some of the larger pieces of ash and alder to the flames, encouraging the fire to build. I enjoyed my evening meal sitting out under the alders while admiring the gate, which was propped against the hedgerow, ready to install. Behind me the wagon's interior was illuminated by lamplight while I sat fireside with the radio for company. By this time of year I was spending the greater part of most evenings outside by the fire, drinking tea or sometimes a drop of cider now that the evenings were lengthening.

After breakfasting the following morning I set about making the hazel rose arch. This would be four feet wide and three feet deep, and sit astride the woven gate, reaching seven feet high at the peak of its curved top. The ground by the hedgerow behind my pitch was reasonably level and a good spot to build the arch, which would be woven on its framework whilst in a horizontal position. The technique I use when making a rose arch is the same as for wattle fencing. The uprights, which will of course be horizontal when the arch is in place, are positioned in the shape of the archway and I can then build the weave in layers to the required depth. By leaving a gap of around six inches between the first row of the weave and the ground, the uprights can be cut once the desired depth has been reached and the archway can then be pulled upright. There are no fastenings used in the build of the rose arch, or for that matter in any wattle work; no nails or staples are needed and it is solely the strength of the weave which binds the structure together. It is necessary, however, to form a hurdle maker's knot when finishing each row of weaving which locks the structure together and ensures that it retains its shape when erected. As each row is finished, the excess hazel is twisted in a cranking motion that separates the fibres of the hazel rod, essentially turning it into rope that can be tightly wrapped around the last upright before then being tucked back within the weave.

I worked steadily all morning, in the shade of the alders, weaving the rods carefully into place, and all the while ensuring that the structure held its shape by bending and stretching the fibres of the uprights as was needed to ensure

APRIL

the arch retained its form and was evenly proportioned. Rolling the hazel rods between the uprights was pleasing work and there was a sense of satisfaction in engaging in this timeless practice, unchanged over millennia. Our very earliest ancestors worked the understorey of the wildwood in the same fashion as we do today, harvesting coppiced rods for a number of building practices. On the wetlands of the Somerset Levels, just a few miles downriver from my pitch here on the Marsh, the ancient remains of wattle trackways have been discovered, which were used to navigate the watery landscape in its Neolithic past, some four and a half thousand years ago.

By the early afternoon, the arch was finished and I was able to cut the uprights and lift the structure clear of the ground. Standing vertically, it now looked considerably bigger than it had done when laid horizontally, and I lifted the gate from its resting place by the hedge to see how the two items sat together. After a little consideration I felt satisfied with the outcome. The herringbone pattern of the woven archway complemented the split hazel of the gate that itself was further enhanced by the green oak framework, and both were products of sustainably managed woodlands.

I was grateful for Tim's help in loading the rose arch on to the pickup, which though not particularly heavy was bulky and awkward to handle on my own. When I arrived at the cottage where I was installing the gate, I was welcomed warmly with a tray laden with tea and biscuits. I looked over the blackthorn hedgerows I had laid that winter, which were now coming into leaf, though a few fading blossoms still hung on. Blackthorn does not grow as quickly as either hawthorn or hazel, but I have found that the regeneration after being laid is often denser than these other species, certainly more so than the hazel. Though hazel regenerates profusely from the stump or coppice stool, it does so only sporadically from the laid stems and so density must be encouraged by layering, pleaching stems before covering them with earth to encourage them to take root and create a new plant. The hawthorn, however, and the blackthorn especially, will send new shoots out all along the length of the newly laid stem, and this was already the case here. Though it was still early in the growing season and just on the cusp of the spring flush of growth that would come in May, tight crimson nodules were emerging from not only the stump but also all around the pleaching cut. The coming weeks would see these erupt skyward, multiplying the density of the existing plants and creating the lush boundary that the customer had so desired. The ever-present robin that seemed always to accompany me, and that now skipped along the binding amongst the thorns, would also be grateful for this enhanced habitat, now fortified and restored.

 I was thankful that the ground was relatively easy digging

APRIL

and I was able to get the gate posts secure and in place within just a couple of hours. The gate then sat neatly between them and, I thought, suited the location well. The rose arch took some manoeuvring and not a little effort to get it into position. It sat astride the new gateway and rested on some flat stones to keep the arch's base from contact with the earth, as this would prolong its lifespan. Finally, I drove two hazel stakes in on each side to support the arch, which would keep it firmly in place, solid and steadfast between the blackthorns.

It was mid-afternoon when I left the village and headed back to the Marsh. Glastonbury Tor and the ancient ruined church that sits atop it rose from the levels that stretched out before me as I took the lanes west. The newly returned swallows swarmed above and around me as I drove up the stone track toward the orchard, and the hedgerows that enclosed it were now abundant with new life as the cold, dark winter months began to fade into memory. Spring was arriving, in all her glory.

10

May

I awoke early, the sounds of the countryside coupled with the lightening sky rousing me from sleep although the time was yet before six. The window above me had been left open throughout the night and the sounds of the May morning permeated the sleeping compartment. I was inclined to remain there for a while and savour the sounds, which were as yet largely uninterrupted by human activity. The hedgerow and alder trees above me throbbed with life, the low humming of bees and other insects on the wing, busying themselves working methodically through the may blossom which hung heavily amongst the hawthorn. The early-morning breeze caused its blossom-laden boughs to sway and the steely thorns lazily brushed the wagon's exterior. I could hear the swallows chattering as their marauding sorties swept effortlessly across the pastures of the Marsh, where the grasses had been left un-grazed and were growing tall in the flurry of spring growth, before they would be harvested for hay in the long June days that lay ahead.

The birds that resided on the Marsh alongside me were already well into their working day, foraging the myriad of invertebrates that dwelt amongst hedgerow and ditches, orchard and hay meadow, in an effort to sustain their demanding broods that lay hidden amongst thorn and bramble. The robin and wren both had nests low in the hedgerow where the lengthening meadow grass grew among the hawthorn's sharp spines. The song thrush had chosen a more elevated position amongst a bramble-smothered elder that grew unchecked beneath the alder trees, which themselves sheltered a pair of blue tits. These familiar little birds had nested in a well-concealed crevice in the tree's fork and now worked tirelessly throughout the day bringing insects back to their waiting offspring.

Peering through the open window, I could see the overgrown hawthorn hedge that ran across the hay meadow beyond the wagon, tracking eastwards from the ancient willow pollards that bordered the track to the cider orchard. It had been neither trimmed nor laid for many decades, and was now a line of mature thorn trees through which bramble and dog rose rambled freely. The briars had not yet managed to overcome the thorns and although the trees were well over a hundred years old, they were still strong and resilient and were currently in full bloom, entirely covered in the thick veil of blossom which gives them their affectionate epithet 'the bride of the hedgerow'. The sun was by now beginning to gain height behind this remnant hedge and its late-spring rays carried the life-giving warmth that fuelled the bustle of activity I could hear around me, the seasonal throb of new life that is so apparent in May, perhaps the most celebrated month in the countryman's calendar.

Eventually I rose and proceeded to make tea on the gas hob. The sunshine poured into the wagon from the east-facing window above the desk, and I opened the door wide; there was barely a breath of wind outside and the gas flame didn't so much as flicker in the still air. I sat outside for a while and lit a fire to heat some water in the enamel saucepan for washing – there was little sense in using the gas for this task on such a calm morning. As I waited for the water to boil I heard what, in recent years, has become an increasingly rare sound in our countryside. From amongst the May blooms of the hawthorn hedge came the unmistakable, two-toned song of the cuckoo. Four or five times in quick succession, its call rang out before the bird paused for half a minute before calling once again.

The cuckoo will not stay on these marshes for long, as it needs only to mate and lay eggs, and will leave the rearing of its offspring to another. The cuckoo will select the nest of a different bird, and will wait until the chosen surrogate is absent before carefully removing one of the host bird's eggs and laying its own in its place. Upon returning, the unsuspecting host will continue to nurture and protect the clutch, completely unaware of the deception. When the young cuckoo hatches it too becomes complicit, despite being no more than a shell-bound embryo when its parent put the plan in motion. The featherless, blind cuckoo chick is compelled to kick out with its legs and stumpy, unformed wings, pushing the other eggs in the clutch out of the nest. The frantic host does not understand what lies behind this behaviour, but its instinct remains to nurture and continue to feed this one remaining offspring even as it fledges, bringing it morsels and encouraging its flight though the fledgling may now be twice its size. Such is the host bird's protective instinct, it will work tirelessly until at last the young cuckoo leaves its exhausted surrogate and prepares to migrate south, following in the path of the adult cuckoos that will have left these shores for Africa several weeks earlier, by the time the hay is cut in late June.

The reed warbler is one of the cuckoo's most favoured hosts. This small bird thrives in the Somerset wetlands and has succeeded in conserving the cuckoo in greater numbers than elsewhere in the countryside, although even here its distinct calling is still rare. Despite the cuckoo's somewhat indolent parenting habits, it remains a bird for which many have a great fondness. Its lucid song will only ever reach the ear when the weather is calm and settled and the countryside basks resplendent in the weeks approaching midsummer. In these kinder months of the year, it seems,

MAY

the cuckoo's deception and idleness are more easily forgiven.

I had been busy. Between regular trips to Hampshire to collect hazel from Earnest, and my wattle work, I had also repaired the post-and-rail fencing around the orchard as well as cutting and processing a good deal of wind-blown timber from around the farm. Though I receive many enquiries for wattle fencing, I take on just a handful of contracts over the spring months. Sourcing hazel in sufficient quantities is difficult and though Earnest has been most accommodating, he is just one man supplying several rural tradesmen and so must limit the demands on him for his own preservation.

Earnest and I have talked at length about the future of the coppice about which he is so passionate. Many lifetimes have been spent working the woods there and he is justifiably fearful that the line of woodsman who have preserved it thus far could end with him. The benefits of working among the woodlands and hedgerows are many, though there can be no doubt that the financial rewards compare poorly to many other trades. The traditional working woodsman will never get rich from his labour and there will be little opportunity to expand his business to a point where he may step back and leave his labours in the hands of others. Realistically, he will have to work among the understorey until poor health or old age renders his tasks impossible. To the right men, however, who see value in tradition and longevity and who look beyond their own lifetime to the continuation of a working rural landscape which they themselves inherited from generations of past countrymen, the preservation of the woodsman's role is of the utmost importance. Despite some adversity, the woodsman will see many benefits in his occupation. He will not be a slave to his work, though it is undoubtedly physically demanding. There are few who work in such close alignment with the seasons and see so intimately the ebb and flow of the year, who work so harmoniously with their environment and directly improve it. Of course, the woodsman's income will never see him living extravagantly by today's standards. To work as a woodsman you must place little value in material gains and see no shame in living simply; you must remember that striving for such superficial ends will benefit neither you nor the environment and will ultimately line the pockets of men who care little for the natural world. It is, in my opinion, a humble but rewarding existence that although not without its hardships also gives back a lot, not only to the land itself, but also to those who partake in its management. I have been much enthused by the renewed interest in hedging of late, seeing the craft increasingly promoted on various platforms, and I hope this will also lead to the

MAY

importance of the coppice woodlands being more widely recognised.

I had by now been on the Marsh for a little more than six weeks and had completed several wattle fences during my stay. A lull in my contract work meant that for the first period in some weeks, I had time on my hands. After some consideration and after speaking with Tim, who informed me that the hay was still several weeks off being cut, I decided to leave the Marsh for a while and travel eastwards, to Dorset. I had lived in this county for many years whilst my children were growing up and intended now to revisit an overstood coppice that lay only a short distance from the village where they were raised. I would be able to cut a good number of hedging stakes from the woodland in preparation for the new hedging season that, though still over three months away, I knew would come around quickly. Both of the girls would be in the village over the summer months and I would be able to see them whilst cutting in the woodland, which was something I very much looked forward to. And so, after packing down the wagon, I pulled away from the Marsh, the old lorry lurching along unceremoniously beneath the willow pollards which were by now in full leaf, before reaching the elm hedgerows that guided me past the farm and onwards through the village.

I headed south-east, beyond Dorchester, the county town of Dorset, and past the ancient hill fort of Maiden Castle, to the village of Puddletown, rechristened 'Weatherbury' by Thomas Hardy in his novel *The Mayor of Casterbridge*. Just over the hill from Puddletown lies a woodland of the same name, where I would be staying for a short time, cutting hazel stakes. I left the main road at the village and took a smaller, unmaintained lane heading south. I was faced almost immediately with a steep ascent that led me into the woods, and easing the wagon's ageing engine into second gear, I coaxed it onwards along the barely navigable lane which was clearly only used by timber lorries and forestry vehicles. Deep potholes and ruts had been filled with loose stone which had since been rinsed away by the heavy spring showers and now lay scattered across the road's cratered surface. I crept forward at a snail's pace until the road eventually improved and began to descend through aged beech which formed a belt around a plantation of Douglas fir. The late-spring sunshine illuminated the lane before me, its soft light diffracted on to the rutted gravel in dappled patterns by the beech's bright young leaves, chlorophyll prisms that filtered the sun's ever-strengthening spring rays.

Eventually, the lane levelled out into a wide turning circle. Before me lay a field of corn enclosed by flowering hawthorn, a brick-and-flint farmhouse just visible on the far side, whilst the woodland carried on to my east and west. The western flank largely comprised mature Douglas fir, all that remained of a crop that had already been thinned several times. These remaining stems were impressive in size, but had no understorey of which to speak, apart from the creeping rhododendron that thrived on the acidic soil beneath them and left little room for any more desirable species. The tough, pioneering birch had gained a foothold here and there, though each silver-barked tree was dwarfed by the towering Douglases. To the east, however, the wood was different, made up to a greater extent of deciduous natives, oak and ash in whose dappled shade grew a mixed understorey of holly, rowan, hazel and hawthorn. Eventually, the oak standards thinned until the wood became composed entirely of hazel before coming to an end perhaps a quarter of a mile from where I now parked. Beyond the edge of the wood was cultivated farmland and the village.

I drove on, leaving the Douglas plantation behind me. On my left now was the hazel wood and on my right the cornfield and white blossom of the hawthorn hedgerow interspersed with mature oaks which had woken from their long winter dormancy, unfurling luminous green leaves above me. Toward the end of the hazel coppice and opposite one of the oaks on the lane was a wide lay-by and a gate into the woodland. It was just beyond the gate that I would make my pitch, under the overhanging stems of the hazel which encroached into the lane. I was not here uninvited; I knew the owner of this woodland, who had purchased it from an estate which at one time had owned great swathes of land in this district, though it had long since been divided into lots and sold. The old manor house still sat beyond the Douglas plantation amongst the farmland to the south, and the family's name was still present in the murals adorning the village church in Puddletown. But with the acreage of the estate much diminished, the family's legacy had by now faded in all but the oldest inhabitants' memories in this quiet corner of Hardy country.

I slept soundly on the first evening beneath the hazel and awoke to a dawn chorus at its peak. Though the volume of birdsong here in the woodland greatly exceeded that of the Marsh, it was nevertheless a gentle alarm. I could hear the chiffchaff in full voice and then the distinct, broken laugh of a woodpecker as it flew across the wood. The ever-present croaking of rooks from the taller trees beyond the hazel provided a backdrop to the mixed tones of the woodland songbirds who were themselves accompanied by the gentle cooing of a pair of wood pigeons in the oak on the other side of the lane. Descending the wagon steps, I drank my morning tea whilst leaning on the gate and casting an eye into the coppice. At a glance I could see that the hazel poles, though growing straight and tall, were thick and heavy and now of a size more valuable to the charcoal burner than the hedgelayer. I didn't doubt, however, that I

MAY

would be able to find some suitable for stakes, enough to justify my time spent foraging. In any case, I was in no rush, and the primary reason for my visit was the company of my daughters. Any stakes which I managed to harvest would be a bonus and the time spent there amongst the bluebells and wild garlic, which both grew abundantly throughout the copse, would be soothing to both mind and spirit.

It was late afternoon when I became aware of footsteps and familiar voices approaching down the lane. Stepping out from the cover of the wood I could see my two daughters, Rosie and Lily, making their way toward my pitch. Lily was leading her liver-coloured young spaniel whom she had named Freda. Though I had spoken to both girls regularly, many weeks had passed since we had last been together and I was eager to see them both. I stood there for a moment watching them as they approached. They were laughing and talking between themselves as they made their way toward me and as I waved, I caught myself in that moment feeling very proud of them. I am lucky to have a good relationship with both girls, but it is their relationship with each other that gives me the greatest pleasure. Despite their differing personalities and interests the two girls are the best of friends and happiest when in each other's company and supporting each other. It is this unwavering adoration that they have for each other, and the strength they find together when navigating a length of bad road, of which I am particularly proud. Life will always throw up hurdles, and knowing that their bond will help them both when these occur provides me with a lot of comfort.

I put the kettle on to boil and Lily let Freda loose in the wood while we caught up. As we talked the dog suddenly flushed a cock pheasant from its cover beneath the hazel, sending the colourful game bird skyward in a flurry of crowing and wing beating which saw it clear the lane before descending safely into the cornfield beyond the hedgerow. After tea and with plenty of daylight still left, we decided to make our way down the lane to where the Dorset Frome passes through the water meadows. The Frome is a typical South Country chalk stream, and its gin-clear waters flow beneath an ancient stone bridge at a place known as Woodsford, where we had spent many Saturdays when the girls were young, searching for bullheads beneath the stream's stones or sitting idly watching the trout rise at the far end of the deep pool beneath the bridge. The dark waters between the stone arches also held pike that grew to not inconsiderable proportions.

The first flush of spring was apparent everywhere now and the cow parsley bellowed out into the lanes from the verges, its white flowers sitting alongside those of stitchwort and may blossom, filling the air with the sweet scent of spring. On our approach to the bridge, the hedgerows had been replaced on either side by wire fencing that was being engulfed by the cow parsley. The white flowers of the boundaries gave way to the yellow of buttercups and dandelions that peppered the meadow beyond. A herd of beef cattle who were grazing there became aware of our presence and, with the curiosity typical of their species, galloped across the meadow and shadowed us for the last hundred yards before the bridge. Every now and again one of their number would break away from the herd in a seemingly uncontrollable moment of excitement at the heady spring evening, kicking their legs and jumping like a young dog when first set free of the lead.

Upon reaching the bridge, we encountered dense clouds of mayfly. Having hatched earlier in the day and rested, they were now fully immersed in their balletic swansong. All along the river they danced silently, carried vertically into the May sky perhaps thirty feet or more on delicate silken wings, then descending feather-like toward the waters below before rising once more. In this way, the male insect attracts a mate, clasping her tightly in mid-air where together they continue this time-honoured choreography, mating on the wing briefly until the male is spent. He then retires to the thick vegetation that grows from the banks of the chalk stream, where he dies, his work now complete. The female will soon join him, but she has one more task before her brief life on the wing also comes to its end. She descends to the water's surface before laying a stream of eggs that eventually hatch into nymphs. These juveniles reside within the sandy silt that collects on the river bed before they too will finally rise, ascending through clear filtered water to the surface, where they will emerge as adults and mate on a May evening, two years from now.

The activity of the mayflies had not gone unnoticed; indeed, for the residents of the water meadows and river, the event had been much anticipated. The birds gorged themselves on the insects, picking them off easily from the dense cloud, and carrying them back to hungry broods who feasted on this seasonal windfall. The swallows and house martins swarmed on the outskirts of the rising columns and picked off stragglers which had strayed from the safety of the main group. This tactic is employed by many of nature's predators; a seal harrying a shoal of herring will take fish separated from the school in the panic; a peregrine will snatch starlings from the peripheries of the flock.

The mayflies were under attack not only from the air but also from beneath the water's surface. Here, waiting in the current, lay the grayling. The Frome has a good population of these fish, a handsome and streamlined chalk stream salmonoid, the male of which has a sail-like fin that makes it resemble the saltwater marlin. The grayling usually feeds closer to the bottom of the stream taking nymphs, insects still in their larval stage, though with such abundance on offer he makes an exception and rises to the surface, gulping hungrily at the spent mayfly lying trapped in the water's film. It is the brown trout, however, the resident game fish which grows to weights exceeding three pounds, that has been most anticipating the mayfly hatching. He will not come across such a bounty at any other time over the next twelve months and so he devours the insects gluttonously. Abandoning his usually cautious demeanour the trout strikes in all directions, taking his fill, before eventually retreating sluggishly to deeper water to digest the meal.

The girls and I walked beside the river for perhaps a half-mile or so, whilst letting the dog run free in front of us. The river was a hive of activity with fish jumping clear of the water, female mallards skipping across weed with ducklings in tow and a mute swan still sitting patiently on eggs that would be hatching any day now. The male swan patrolled vigilantly and watched us, and the dog particularly, who after a brief glance paid the birds no mind. Lily had spent a lot of time on Freda's training and the spaniel obeyed her commands impeccably, responding to a whistle or word. An ill-trained spaniel is a hindrance to owners and wildlife alike on a walk, but with adequate training it is a favourite breed for country people whether it is a working dog or simply a companion.

It was almost dark when I bid farewell to the girls and they headed back into the village while I made my way back down the lane to the woods and my waiting wagon. The sun had sunk below the towering Douglas firs away to the west and the first stars were now visible in the sky above me. On entering the wagon I realised how tired I felt, and lit only one candle to illuminate the living space just briefly whilst I undressed and prepared for bed, making sure it was extinguished completely before retiring. The wood pigeons in the oak cooed softly to each other as I quickly drifted to sleep.

Once again, I was awake with the birds. At this time of year it is almost impossible not to be, living as I do amongst woodland and hedgerows. As I drank my morning tea and gradually came to, I opened the top of the stable door to let the air in, but left the bottom of the door closed to provide me with some privacy from passing cars or dog walkers from the village who often came this way. I had spoken to many of the villagers while parked up here before, and had been greeted warmly by most, though I am sometimes still met with suspicion and on occasion even hostility, as is sadly not unusual when living in what many consider to be an unconventional fashion. In fact, Britain has supported an itinerant population for many centuries, and agriculture has historically relied on this population throughout the year and particularly over harvest time. Of these itinerants perhaps the most well-known are the Gypsy or Traveller people, whose presence was first recorded on these isles in the sixteenth century. The Romany had roots stretching back to India, from where their ancestors had first migrated; upon arrival on the British isles they were mistakenly thought to have heralded from Egypt and were initially referred to as Egyptians, which was quickly shortened to 'Gyptians' and finally to 'Gypsies' in the years following their arrival.

Though widely mistrusted and persecuted from the outset, gradually the Romany and other travelling people became interwoven into the fabric of British rural life and went on to contribute in no small amount to the farming calendar. When not engaged in agricultural work, of which hedgelaying often played a significant part, the Romany spent much time absorbed in craft. They worked with hazel and other woods gathered from the coppice to create clothes pegs, baskets, walking sticks, umbrellas and brushes which could be hawked door to door in towns and villages close to their camps. They did, however, live on the fringes of society, much as they still do today, pursuing traditions and a way of living largely unknown to the settled communities around them, while also speaking a language which was understood by few. As a result they were often viewed with a deep suspicion, as is so often the case with anything that people don't fully understand or which, upon observing only briefly, appears to conflict with the values and traditions deeply instilled into their own ways of living. And so I still find today, even here, where I reside roadside in the wagon, cutting hazel for hedging stakes by kind permission of a landowner whom I have known for many years, though I am usually greeted warmly and often with a curiosity that leads into conversation about the craft that has taken me on this journey, some will still hurry past or mutter disapprovingly under their breath, seemingly offended at my lifestyle, or at least their interpretation of it. For my part, I am often amused, perhaps slightly bewildered, that in apparently enlightened times living above a set of wheels can still condemn one to such harsh judgement by seemingly learned and intellectual people, but apparently it has ever been so.

I had spent the morning cutting stakes and by now had two hundred or more straight hazel poles of a suitable diameter, cut to roughly five foot six, my preferred height for hedging stakes, stacked under a mature beech which lay on the edge of the copse and just a short distance from the wagon. After stopping for a quick lunch of soup and toast I retrieved my axe from the locker beneath the wagon before heading back to the beech tree, where I intended to point the stakes I'd collected that morning. I had only just begun when a voice called out, interrupting my work. I barely had a chance to reply before Freda was excitedly running about me, whining and barking in the kind of over-joyous greeting one only receives from a dog. Eleanor, the mother of my girls, was at the gate; she had been enjoying a walk in the fair weather we were experiencing and had decided to stop in on me. She greeted me warmly, with a broad smile that I returned, before offering me some freshly baked biscuits, still warm, which she pulled from the satchel she carried. She had often baked when the children were young and I was pleased to see she had kept up the habit, and remembered my fondness for her cooking. Though our relationship had seen both affection and conflict, past hurts had been forgiven on both sides and the bond that had held us together over the last couple of decades, though now changed, remained unsevered.

This was an unexpected visit, but a welcome one, so I left my work and we headed back to the wagon for tea and a chat. She filled me in on life in the village and what had changed since my last visit. I had spent a good part of my life in this village, had raised both girls to the best of my ability here, and at one time had presumed that this was where I would grow old. I had first moved to Puddletown when I was just twenty-one, with Eleanor, four years my senior and at that point six months pregnant with our first daughter. We had been working on a fruit farm in Herefordshire over the summer, engaged in seasonal agricultural work of the type now more often carried out by Eastern European labour, though in those days many farms were still largely reliant on the local or itinerant workforce. At the time, we were living in an antiquated mobile shop that had been converted into a living van, and with the baby due at the end of November we took the decision between us to settle, if only for a short time, in West Dorset.

Eleanor's mother had bought an old cottage there some twenty years previously to provide accommodation for visiting friends or relatives, but this was a rare occurrence and the cottage lay empty and unmaintained for the greater part of the year. It had not been modernised in any way, and was heated by a woodstove and open fire, though the single-glazed cottage windows did little to hold the warmth. The walls were made of a crumbling cob that would eventually warm through, but we had to keep the fire burning constantly throughout the winter months to maintain any sort of comfortable environment. Despite the lack of home comforts, the old cottage certainly provided more suitable conditions than the van which had served as our home for the previous eighteen months, and with both the baby and winter imminent we were grateful for the improved circumstances.

During the late summer of 1997 we moved into the cottage, exchanged the old living van for an estate car better suited to parental duties, and embarked on a more settled existence. Within three months of the move Rosie was born at the county hospital in Dorchester and I began working the woods with old Bill Bugler, earning little, but enough to pay the rent and keep the wood store full. Those early days were a struggle, though I look back on them now with great fondness. They seem not long ago, but the years have moved quickly and Rosie has now matured into a young woman of whom I am very proud, and is forging her own path through life, while her sister Lily, four years her junior, is also following suit. Both girls now live away from the cottage for the greater part of the year but retain a love of the fields, streams and woodlands where they were fortunate enough to grow up, and Lily is studying practical land management and embracing the skills of the coppice worker, hedger and charcoal burner. Those early days in the cottage are now over a quarter of a century ago, but the trials that Eleanor and I faced together raising the girls during those times bond us still. The love between us did evolve as time passed, and eventually led us on to different roads, but it was not lost altogether, and we are still able to laugh and reminisce on times gone by and, of course, celebrate our daughters' successes.

I left Puddletown for good some years ago. By that time, the girls had grown up and were embarking on their own lives and my work, too, had become busier, with an increased interest in the hedger's craft leading me westward, away from this place. Though that chapter of my life is now closed, it is one I remain fond of, and which I occasionally revisit. My life's journey, like most, has contained many hard-learned lessons, along with some regret and hurt, but I feel fortunate that it has also seen much love and laughter and has rewarded me with memories that bring me great joy. I believe success should not be measured in the accumulation of material wealth, but in our relationships with each other, with our community and with the land. When the winter of our lifetime comes it is these relationships that bring comfort and that, ultimately, will be our legacy.

After several hours Eleanor returned to the old cottage where she still lived and where the girls were staying over these summer months. I made my way back into the copse and resumed my work under the beech, pointing the stakes by bringing the sharpened axe down on the butt end of each pole in four quick blows to leave a sharp point which would see the stakes driven home easily between pleachers. As the afternoon turned to evening and the light in the wood began to retreat, I felt thoughtful. The birdsong began to increase in volume, regularly interrupted by the sharp blows of my axe falling upon the hazel poles, the loud crack of steel on wood echoing out across the woodland. I felt, not sad, but reflective, after Eleanor's visit, aware of the passing of time more keenly than usual.

The lane outside the wood was quiet and I thought of the people in the village all now engaged in their nightly routines, settling in for the evening after the day's activity. With only the birds for company I decided to light a fire with the hazel chips and dry wood that now lay scattered under the beech. First, I retrieved a bottle of cider from the wagon. I find a drop of cider soothing on such evenings, when the hours spent in solitude can see a feeling of

MAY

melancholy descend. I value my relationships greatly, both with the people I have grown close to and with the environment in which I live. I endeavour to take neither for granted and seek to nurture and preserve them both. When separated from the individuals who help sustain us through the various hardships of life, we can find the hope and resilience we need in the ebb and flow of nature's calendar. We all seek consistency and reliability, and often descend into anxiety without their reassuring structure. During the month of May, perhaps more than any other, nature shows us this unwavering consistency and reliability in its timeless cycle of rebirth: the hypnotic dance of the mayfly, the blossoming of the hawthorn, the swallows streaking wildly across the water meadows, events that have repeated for millennia, alongside the humans who have also worked in unison with nature's year. We evolved within the cycles of nature's calendar, and it is therefore in the consistency of nature's rhythms that we thrive. Our ancestors, both ancient and more recent, rightly cherished this relationship and nurtured it appropriately, as one must nurture any relationship if you wish to see it thrive. If we take this relationship for granted, however, neglecting it and, worse, exploiting it continuously for our own ends and habitually disregarding its beauty and vulnerability, we will lose it.

I had by now gathered a good number of stakes, more in fact than I had thought possible when first glancing into the wood. Two substantial stacks of them now sat under the beech, approaching eight hundred in number. I had tied them in bundles of ten and stacked them in layers of one hundred, four high with each layer in opposing directions. I was pleased with my harvest; the white wood of the freshly pointed stakes shone and hazel chips were scattered widely about an old stump that I had been using for a seat whilst I worked. May was now drawing to an end and June would soon be upon us. I still had one more fence to weave in a hamlet that lay just outside of Glastonbury on the Somerset Levels, and would use the last of the hazel rods I had collected from Earnest, which were still propped vertically in the shade of the alders on the Marsh and would be malleable for only a few more weeks. Though hazel can be gathered and worked all year round, the rods harvested in the winter months are best for wattle work. Hazel cut during the colder months of the year when the plant is dormant will have greater longevity than the summer-cut rods, which are full of sugary sap. Though there is little difference in the way the hazel can be worked, the summer-cut wood will decline more quickly and for this reason is usually avoided. My upcoming wattle job would be my last of the season, and the following months would be spent employed in various tasks around Marsh Farm.

Before I moved on to this last job, however, I was craving

MAY

an afternoon spent wholly in leisure, and decided to see if I could relieve the Frome of one of its trout. During the evening I had spent by the river with the girls I had spotted several good fish and had noted where they lay, hanging in the current above the smaller fish and taking the lion's share of the insects hovering above the water's surface. Tucked away at the back of the wardrobe in the wagon, in an old cloth sleeve, was a fly rod, now little used, along with a small game bag that contained a reel holding a light trout line of the type used on these smaller waters. There was also a small wooden box containing a handful of flies designed to tempt a trout, impersonating various up-winged insects and nymphs such as March browns and mayfly, dragonfly larvae and daddy-long-legs, different insects for different times of the year. Some were designed to sit low, close to the silt on the river bed, whilst others would float gently on the water's surface film, but all were designed to deceive the wily trout into taking this morsel, the hunter becoming the hunted.

As afternoon turned to evening, I made my way once more down the lanes to the water meadows at Woodsford. The evening was quiet and the sun hung just above the horizon. The chalk stream's waters flowed like thick oil, rippling and folding silently as they carved their way across the meadows. The low sun did not illuminate the water and my figure cast no shadow upon it. A reed warbler could be heard, but remained unseen from the cover of the far bank.

The trout, however, readily disclosed his location as he rose from the dark water to sip the last of the mayfly fruitlessly struggling on the water's surface. I watched him for perhaps ten minutes, all the while surveying the river carefully and becoming accustomed to the trout's habits, tuning in to his environment as best as I was able, excluding other distractions from my mind, focused completely on him.

Quietly I removed the rod from its sleeve, placing the two parts together and, attaching the reel in its seat, drawing the ivory-coloured fly line through the rod's eyes before tying a clear leader of nylon to its end, where I would attach the fly. I selected a mayfly pattern from the box, which was made from cotton and feather and had an uncanny resemblance to the insects that drifted past me in the air. From the wide bend of the river just beyond my spot I could hear the grayling surfacing to feast on the struggling mayfly which had avoided the deeper water of the meander only to be enthusiastically picked off by the smaller fish who lurked in the shallows. I pinched some silty chalk mud from the bank of the stream and then rolled the nylon between my fingers; this would cause it to sink and present the fly in as natural a fashion as possible. Although the trout had been spoiled for choice over the last few days and might have grown complacent, he would not necessarily be fooled easily.

From the nearside of the curve I cast my line, flinging it three times low behind me across the meadow and then propelling it forward with a snap of the wrist, the line gaining distance with each whip of the rod. Once sufficient line was airborne, I brought it to an abrupt stop by placing a finger upon it at the point where it left the reel, this causing the mud-weighted nylon to unfurl lightly and drop the fly gently on to the surface film. Unhindered by the current, the fly then drifted naturally through the trout's field of vision, tempting him. Three times I made the cast and though the fly glided provocatively through his lie, he remained steadfast. Doubt crept in, but I continued to wait and watch, and the trout rose intermittently. On the fourth cast and with the light now fading, the fly landed perfectly just five yards from where the water's surface was pockmarked by his previous rises. I held my breath. The trout rose – and engulfed the fly. The ivory fly line jabbed quickly sideways and I drew the rod sharply upwards whilst holding the line with my left hand, and maintaining close contact with the fish who now shook his head frantically whilst coursing at great speed through the deep pool of the river bend. Several times he leaped free of the water, shaking his head in an attempt to free himself from his tether, his struggles sending other fish fleeing into weed and reed beds. Eventually, like the mayfly which still passed on the current, the trout too was spent; he came gently to my hand and I lifted him carefully from the water.

He was a creature of great beauty, his smooth, buttery-coloured belly giving way to a darker green-brown back that was pricked sporadically with handsome, deep red spots of differing sizes. He was in every sense a predator, as handsome in his own way as any cheetah and just as they are, perfectly evolved for his environment. For a while I simply admired him, cradled in my hands in the still, cool water of the river's edge. I had planned to have him for the pot, as fresh trout is one of the greatest of nature's gifts, but I found myself reluctant to take his life. Apart from the energy expended in his struggle he bore no injury — the small barbless hook I used had done little but to tether him to me during the fight — and I knew he could recover fully with a little rest and be feeding on the mayfly again come the following evening. So I stayed crouched by the water's edge and allowed him to recover. He took just one or two short minutes resting in my cupped hands, the chalky waters pulsing through his gills and filling his muscles once more with oxygen, his strength returning all the while. Then his broad, flat tail began to wave more determinedly before he suddenly darted free from my palms, returning unharmed to the clear, chalk-filtered waters of the Dorset Frome.

11

June

A monsoon descended. The rain fell vertically from the blackened June sky in heavy drops that exploded violently when they hit the cracked tar of the lane. Water ran down the narrow road from the village in a torrent and was stained a cloudy white from the chalk of the cornfield as it rushed beneath the field gate at the top of the slope. From the open door of the wagon I heard the low, loud rumble of thunder that passed overhead in a stereophonic wave from west to east, and I kept a close eye on the turbulent sky for the lightning that would inevitably follow. Moments later, beyond the cornfield to the south, the sky unloaded its charge and the heavy midsummer air was fractured momentarily by electric light which descended instantaneously and at random. The storm tracked eastwards, violently haemorrhaging energy. From the confines of the wagon I listened to it pass overhead, the rain lashing the sides, droplets hitting the thin tin sheets of the roof like a hail of bullets. Despite the ferocity of the storm I left the top of the stable door open so that I could enjoy the spectacle unfolding outdoors. I looked down the lane to the west, beyond the hazel and the cornfield, toward the Douglas plantation in the distance where the evergreen firs were dancing in unison, absorbing the energy of the storm and bending to its will.

It was above the farmland to the south that the sky first began to lighten. This summer storm was not like the heavy winter gloom that descends and blankets the sky for days on end, blocking the sun's rays indefinitely from the cold earth. This June storm was wild, untamed and galloped unbridled across the sky, then was gone, passing quickly. It had silenced the birds and animals of the fields and woodland, who only tentatively began to find their voices as the storm disappeared to the east. With the worst over, I stepped outside, and found water still rushing down the lane in the storm's wake. Within minutes, though, the hot June sunshine reappeared and lit up the cornfield before me. The shell-shocked wood pigeons cooed nervously, comforting each other from the nearby oak. Both pigeons looked bedraggled; the female had sat resolute on her nest throughout the storm, exposed to the downpour but instinctively protecting her vulnerable offspring and ensuring their survival.

With the passing of the storm my time had come to leave this place and I set about packing my belongings down in the wagon. The year was moving on and Midsummer's

JUNE

Day was now but a fortnight distant. I had still a wattle fence to install on the Levels and then, weather permitting, haymaking would be upon us.

I noticed the may blossom had all but disappeared as I pulled on to the Marsh, and the hedgerows retained only a light peppering of the fading white flowers. The elder that grew under the alder trees in which the song thrush was nesting was now picking up the baton, and its creamy, off-white blossom was creeping throughout the hedgerow that separated the cornfield from the pasture. It was maintained only by an annual trim, though many of the elder trees had escaped the flail altogether and now grew tall under the alders and ash where the hedge cutter had been unable to reach. The plant now dominated great lengths of the hedgerow and largely outnumbered the hawthorn of which it was originally constructed. The old growth of some of the elders had died back and become brittle and it now provided a climbing frame for the brambles that were just coming in to flower. The elder's dead limbs, wrinkled and bent, protruded from the scrub like witches' fingers, parting the hooked tendrils of the briars.

The elder tree is a welcome addition to a hedgerow. Its blossom sustains pollinators and is sought by foraging country people who make cordial from the flowers which, when served chilled, provides a welcome antidote to the labours of hay time. The elderberry, too, is made into any number of hedgerow preserves or fermented to make a country wine. However, like the bramble, when elder becomes the dominant species it is a sign that the hedgerow is in decline. Elder seeds are often introduced into the hedgerow by the birds who gorge on its fruits in the autumn months. Once the new plant gains a foothold, the elder outcompetes the other species to be found in the hedge, broadening its form and growing at a rate which the thorns are unable to match, in a fashion similar to the willow. The elder will, however, often overstretch itself and tends to die back. Although new leaders will emerge, ultimately this behaviour will lead to gaps forming in the boundary which are exploited by the creeping brambles, and both elder and bramble will then silently seek to overwhelm the thorns and outmanoeuvre the other plants of the hedgerow.

The hedger has always kept a close eye on the elder with all of the plants usually being cut out of the boundary when laying. They would always return, but under the hedger's watch they were contained and the boundary's integrity preserved. With the majority of hedgerows now maintained by mechanical trimming alone, the elder has seen many gains in its ongoing battle for dominance. If walking in the countryside in June and looking upon hedgerows that have obviously been trimmed during the last winter, you may well see a shrub that has grown taller by several feet than the surrounding plants; if not the searching leaders of the willow, these stems will belong to the elder tree.

JUNE

Though the hawthorn flowers were all but gone, the hedgerow continued to throb with life. The delicate dog rose was blooming alongside the elder, each flower having just five delicate pink petals that fade to white at the centre; from here the flower's yellow pistils radiate outward like the June sun. The flower attracts bees and other pollinators in such numbers that the hedgerow was vibrating with activity as the insects busied themselves between elder flower, bramble blossom and dog rose, and even foraged amongst the diminishing may blossom. The pasture too had changed in the few weeks I had been away. From the willow pollards to the orchard and out toward the old remnant hawthorn in the distance, the grass had grown long and thick and now rippled as it caught the warm breeze. I knew Tim would be keeping a close eye on the weather by now, listening carefully to the forecast and waiting on a clear window that could see him mow, dry and compress this summer grass into small bales that we would then manhandle into the Dutch barn during the cooler evenings. There would be in the region of five thousand bales to stack over the period, harvested from several different meadows on the acreage and varying in quality. They would provide winter feed for the cattle and sheep on the farm and he would sell a good number to local equine yards, so the successful harvest of the hay was of great importance to the farm.

The weather had settled by Midsummer's Day. As I drove up the orchard track back to the wagon, the mud of previous weeks had been replaced by dust that was now kicked up by the farm traffic and carried off in the warm summer breeze. Tim had mowed the hay on the Marsh two days previously and had left it to settle, and was now working the field with the 'hay bob' fastened to the rear of the tractor, flicking the drying grass into loose rows. This process turned the hay, ensuring the summer rays could reach every part of it, and he would do this several times before he baled the crop, making sure it was dried thoroughly. Damp hay heats up rapidly when compressed into tight bales and stacked closely in a barn, and Tim knew more than one man who had lost his sheds and hay harvest due to fire caused by the damp hay combusting, as the flames will spread rapidly amongst the dry grasses in a barn's confines.

Back at the wagon, I set about making tea. I caught Tim's eye and made a 'T' with my hands to let him know there was a cup waiting, and he put his thumb up enthusiastically. Tim was not in a modern, air-conditioned machine of the type used for silaging on the big dairy operations in the district – he was hanging out the rear window of an antiquated John Deere whilst grappling with the steering wheel, looking for the greater part backwards at the condition of the hay behind him, and only occasionally casting an eye forward. Tim was born on this farm, as was his father and his father's father, and as I watched him work I wouldn't have bet against him being able to navigate this field with his eyes closed.

He pulled up next to me before the tea had gotten cold. 'What's on then, Paul?' he asked after taking the tea. 'Youm finished all your hazel work by now?' I told him I had, the final rods had been woven in the last few days and I could now relax a bit. 'Not just yet er can't,' he replied, smiling. Tim told me that when he was a boy his father had little trouble in mustering help during hay time, and the men would work mob-handed to get the bales stacked under cover before the rain came. In those days, Tim had told me, there were plenty more bales to contend with. Once the hay was safely in, Tim's father would lead the men to the pub in the village where they could drink their fill; the Taunton cider pump was always kept well primed at hay time. This arrangement had worked well for many years, but things had changed in more recent times. 'Young'uns in the village want no part in it these days,' he told me. 'All 'em want now is to sit in a tractor with the radio on, theym dodges hard work if 'em can.' He was right too, there was no shortage of youth queuing up to drive the modern machines, the huge rigs that now shadowed the foragers and combines run by the big agricultural contractors and estates, but they were scarce when the more 'hands on' operation of stacking the bales was being carried out. Tim, however, always found a way. He was well liked in the district and very involved in its activities, and so managed to muster the labour needed to get the hay in, even if it did take a little longer than it had in those past years.

Tim picked up a handful of the drying grass, which had already lost its bright green colour as it baked, though upon folding several of the stems at their knuckles, the dry sleeve of the sheath cracked to reveal a still sappy and moist interior. Like the hazel and other trees that I work with, the grass at this time of year is full of sugars and moisture. This sap, which causes the hazel to decline quickly when cut in the summer months, would in this case be preserved in the grass by drying it, and would sustain Tim's livestock throughout the winter. Summer's bounty served to the cattle in the sheds on the cold, dark winter mornings.

For the better part of a week the hay lay cut and drying on the Marsh. Tim turned it several times with the hay bob, fluffing and raking the loose rows during the heat of the day. The hay bob consisted of two spinning wheels with steel fingers to turn the hay, flicking it between the rotating prongs to ensure thorough and even drying. Now that it was ready, two paddles had been fastened to the rear of the hay bob and these would channel the dried grass into neat rows that could be processed by the baler. The baler, like the tractor, was an antiquated machine and rarely succeeded in baling the whole hay crop on the farm without slipping

JUNE

a belt or throwing a bearing. A number of visits to the farm workshop would no doubt be required, hunting through straw- and dust-covered tools and spare parts, but Tim knew this machine well and had nursed it through many summers, so inevitably saw the machine repaired in quick time.

It was a warm afternoon when Tim arrived with the baler. I had been watching the swallows streaking above the cut hay and feasting on the insects that had strayed from the safety of the hedgerow and now hovered beneath open blue skies. The baling process was remarkably quick. Tim drove along the neat rows in the old John Deere while the tired baling machine chugged along behind him. Once it had compacted the hay into a tight bale, the machine then held the bales in a steel frame at the back with a sprung gate that would open once its full capacity of eight had been reached. In this way the bales were left tidily in small groups around the field, ready to be loaded by tractor on to the hay trailers before being towed along the dusty track, under the old willow pollards and back to the barns in the farmyard.

Jack Popham, Tim's son, arrived shortly after the baling had been completed and was driving another of the farm's tractors that was coupled up to a heavy hay trailer, one of several owned by the farm. The one Jack had in tow was the largest of the fleet; it had been converted from an articulated lorry trailer and was a good forty feet in length. The bed of the aged trailer had been patched by various planks and boards and though now unsuitable for use on the highway served perfectly for traversing the tracks and fields of the farm at hay time. Fully loaded it would hold upwards of five hundred bales, and unloading its contents into the barn would be hard going.

Early the following morning I breakfasted outside by the fire. Although it was just after seven, the heat of the coming day was already apparent and I was comfortable in short sleeves, drinking tea beside the hedgerow. The hawthorns had now discarded the very last of the may blossom and were a deep green in colour. The bales lay tidily in groups of eight in the field and the hay trailer was parked just a little way down the track, awaiting loading. It was mid-morning by the time Jack arrived in the tractor with the flat eight-loader attached to the front of the machine. This attachment would enable him to pick up each group of bales in one, before bringing them to the waiting trailer. Once delivered, however, the bales would have to be re-arranged on the trailer by hand to ensure that they were knitted together tightly and would not fall when hauling the load back to the farm. This was my first task of hay time.

I walked down the dusty road alongside the blossoming elder to meet Jack, who resembled his father a great deal in both his features and his work ethic, and had inherited not only the farming instinct, but also the good nature that ran throughout the family. Jack would be running this farm one day, and like his father knew every inch of the ground

JUNE

here. With much to do, we chatted only briefly before setting to work. Jack first collected the bales closest to the trailer, and this meant that I had to work swiftly to ensure I didn't hold him up. He lifted each group of bales with the flat eight before dropping them squarely on the deck of the trailer and leaving me to arrange them, alternating between length- and width-ways to ensure the load would rise securely. After what seemed only a short time, we were already eight bales high and I was pleased when Jack stopped to throw me a bottle of water. The heat of the day, coupled with the demanding work, meant my vest was already soaked with sweat and hay was sticking to my arms and face. The sweat, sun and hay dust had dried my throat and I took a long draught of the cool water before pouring a good measure over the top of my head to clear the stinging sweat from my eyes and cool me down.

It was yet not midday, and there were still several more trailers to load. At nine bales high the loader could reach no higher and, in any case, the field here was all but cleared. Between us we decided that I could run the last of the bales to the farm in my pickup. Jack hitched up, lifting the fully laden trailer on which I was still seated on to the tractor's towbar, and leaving me in my elevated position to enjoy the view across the farm and the breeze on my face as we made our way slowly across the Marsh and toward the yard.

It was gone nine in the evening when I walked back

down the track toward the orchard. The swallows were still on the wing and danced and chattered above me, while all around us the other birds of the country sang from the hedgerows and trees. The heron crossed the sky in front of me and acknowledged my presence with a squawk before descending beyond the orchard. The night was warm, the type of evening that can see a man forget a good deal of his concerns or anxieties and leave him free to revel wholly in this season. Affairs other than the turning wheel of the year and its timeless continuity are all but forgotten when walking close to the land and surrounded by this seasonal beauty. If only for a brief moment, the mind finds welcome relief from the heightened demands which often press upon us.

The day had been busy, but the orchard was quiet. I lit a fire and placed a saucepan full of water amongst the flames to warm through while I retrieved a deep enamel bowl from inside the wagon to use as a washbasin. As the light faded and the day cooled, I washed beside the hedgerow, cleaning the hay dust and sweat from my body before finally dousing myself completely with the remaining warm water from the saucepan and pulling on a clean shirt. Then I sat on the steps of the wagon. The sun had by this time retreated fully in the west and the clear blue of the day was replaced by the blue-black midsummer night. The moon, which rose now above the aged apple trees in the orchard, illuminated the cut hayfield and hawthorns beyond and was accompanied by countless stars though the largest and brightest, which lay just off her port side, was in fact a planet, Jupiter.

My mind began to wander, as is so often the case when contemplating the heavens, but my eye was swiftly drawn back to our own planet by a darting and erratic shadow that chased above me momentarily before disappearing into the night. Now that it had caught my attention, I could see that there were several of these creatures; they were swallow-like in their flight, but the swallows had long since left the darkening sky. The air above the hedgerow and hay meadow now belonged to the pipistrelles. These little bats are perfectly designed for night-time forays, light and agile, making sharp turns and dives guided not by sight but by sound alone. Their manoeuvres were faultless, and I was transfixed by the aerial dogfight playing out before me. The moths of the meadow had been drawn hypnotically to the light of my candles and oil lamp, seemingly unable to resist the flickering yellow flames despite the bright summer moon which now climbed ever higher above the orchard. The tiny bats, which weigh little more than a pound coin, gorged on these preoccupied invertebrates and feasted in abandon beneath the alders. They flew above me and at times came so close I would flinch, thinking a collision inevitable, though they always turned at the last moment.

The pipistrelles were as adapted to hunting in this environment as the trout was in the chalk stream, the heron stalking the ditches or the fox who now patrolled under the hawthorns away across the Marsh. All of these animals had evolved wholly in sync with their environments and so thrived there. Surely, I thought as I observed the bats, it is only we humans who have strayed so willingly from the environment to which we were once so finely attuned; we have been lured, like the moths who now flew so willingly to the flame of the oil lamp, down a path that so often sees our spirits wither and die. It is amongst nature, in the meadows and woodlands, streams and hills, that we are closest to the environment we were designed for, and it is when living in a condition that separates us from nature that, in my opinion, we so often suffer. I grew tired at this point, my thoughts made heavy by the warm night and the fire, which now consisted only of fading embers. I retreated to the wagon, casting out several moths who fluttered frantically across the ceiling to run the gauntlet of bats outside, before climbing into bed and descending into a deep sleep.

I drove my pickup down to the farm the next morning after breakfast. It was another warm day and though there were still bales in the fields to be collected, all the hay trailers on the farm were now full. Amongst the fleet currently awaiting unloading in the yard were two smaller trailers, each of them capable of holding between three and four hundred bales, and we decided to unload these first, as this would free enough space to collect the remaining bales that still lay in the fields. The Dutch barn was relatively empty, with just a few bales left over from last year's harvest, which made our task straightforward. Jack hitched up a fully laden trailer and drew it as close to the barn as he could, alongside an empty bay which he had first lined with a bed of pallets; as in the woodshed in Dorset, this would prevent the shed's contents from drawing up moisture from the ground.

The correct stacking of the bales in the shed was an important process, both to make sure that the entire haul would fit but also to make sure they then remained firmly in place. These bales would be used throughout the year, and an avalanche of hay on to the cold, wet and muddy ground when feeding cattle on a dark winter's morning could be avoided if care were taken at this stage and the stacking made secure. I had the easier job of throwing the bales down, and Jack could then stack them in the pattern his father had taught him when he was still a schoolboy. I found a ladder propped against one of the sheds, climbed to the top of the hay trailer and pulled on my old hedging gloves. Though they were by now torn and frayed they offered some protection from the tight bailer twine which cut into my hands, despite the callouses from many hours on the chainsaw. Both Jack and Tim made do without gloves, preferring to clasp a fistful of hay before lifting each bale instead. It was a big job, and took several hours, and we were relieved when the old trailer's patched floor came into view and the final layer of bales had been stacked away.

During the heat of the day we recovered the last of the bales from the meadows. Tim had sold several trailerloads directly off the field this year, and a neighbouring farmer had arrived with his tractor and now worked alongside us to fill his own trailer before towing it back to his barn. This reduced the work burden on us at Marsh Farm, though there were still six fully laden trailers to be stacked away by hand. The window of good weather had already ensured that the harvest was safely dried and back to the yard, and with the hay now off the ground the urgency of the task had lessened to some extent and could be carried out over the next few days. The trailers would be in use again in August to harvest the corn that grew in the field beyond my wagon and in various other locations around the farm. The straw that came as a byproduct of the grain would be baled by a local contractor and used as bedding for the cattle in the sheds over the winter months. These straw bales were much larger than the hay bales, and would be handled by tractor, stacked neatly away in the old straw shed or left under covers on the back of the hay trailers.

Despite working at the farm on both the land work and the lorry deliveries, Jack was also employed several days of the week at a tractor dealership in the town. This enabled him to continue working on the family farm and supporting it through what are increasingly lean times for small family

JUNE

farms. Many have already been swallowed up by the big agricultural operations that now dominate the industry, but Marsh Farm had fared better, and had succeeded in diversifying their operation without losing their identity or culture in the process. Proud and strong farming people who are intimidated by little, the Pophams continued forging their own path despite considerable bureaucratic interference. The farm's ceaseless struggle simply to survive reminded me of the plight of the birds and animals of the marshes and levels, who themselves have seen much disruption though they have been less well equipped to adapt. Our relentless march forward has left many casualties in its wake, and I believe that the loss of many small traditional farms and the methods they employed to work the land sympathetically should be counted alongside the farmland songbirds and other species who have also been lost.

Jack was usually home from work by six o'clock, by which time a good deal of the heat had left the day, and so we worked into the evenings, unloading the hay bales methodically. We had already stacked a good deal of the harvest and the Dutch barn was gradually filling, although the largest trailer still stood waiting to be unloaded, a hay mountain which rose between the sheds in the yard and reminded me every time I passed it of the task that lay ahead. Jack had commandeered some colleagues from the tractor dealership to help and, coupled with Tim and me, this meant that on the evening when we came to tackle the trailer there were five of us present. The hay already rose to fill three quarters of the barn and the bales now needed to go across and up rather than just being thrown down, and for this task there was an invaluable tool. From within the old straw shed Tim wheeled out an antiquated conveyor. It was mounted on a rusting steel frame and its ancient engine was started by a tattered pull cord which had long since lost its plastic toggle and now used a piece of smooth elm wood instead; make do and mend has never gone out of fashion on the vast majority of West Country farms. The starting cord was wrapped and knotted around the piece of elm, and was somewhat frayed itself, but did the job. Once running, the engine rotated a cog which in turn powered the belts that moved the hardwood slats on which the bales would be placed smoothly along the length of the conveyor.

The conveyor was set up alongside the barn and the laden trailer was drawn close to it. The two men from the tractor dealership climbed atop the trailer to pass the bales down to Tim, who would ensure they sat correctly on the conveyor belt while also keeping an eye on the engine to ensure there was no overheating or technical issues. The bales would then travel up the belt to the top of the Dutch barn where myself and Jack would be positioned. I would lift each bale from the conveyor before throwing it to Jack who would ensure that the stacking was continued in the proper fashion right to the top of the arched tin roof of the barn so that

JUNE

the entire haul could be accommodated. With the engine started after a little difficulty, and the conveyor in motion, Jack and I climbed the wooden ladder to the top of barn. Once stepping from the ladder, it was immediately apparent that, sandwiched as we were between the newly stacked bales and the tin roof sheets, the temperature was oppressive. The still air was hot and dusty and both Jack and I had broken sweat before the first bale even arrived.

The bales were loaded two at a time on to the conveyor. Any more than this would see the belts begin to slip and the engine working overly hard and so we started slowly and soon entered into a sustainable rhythm. Jack set about building the back wall first, putting each bale carefully in place and succeeding eventually in completely enclosing us. He then brought the bales closer towards me, gradually filling the remaining space and working from the rear of the barn towards the front. By the time a third of the trailer had been unloaded, the old conveyor was becoming hot under the strain, and the sweet aroma of the meadow hay was now infused with the unmistakable scent of hot oil. Tim took the decision to rest for ten minutes which gave both us, and the engine, a little time to recuperate. His wife Susanne joined us in the yard, emerging from the calf shed carrying cold drinks and ice lollies.

Susanne was herself a farmer's daughter and had helped with the hay harvest many times, both here and on her parents' farm when growing up, and although now spared from the labour here, she could appreciate the physical exertion required. Sitting there chatting with the family and the boys from Jack's work the conversation was timeless. Tim talked of farms who were yet to mow – 'er's left it late I reckon' – or the price the lambs were fetching at the Sedgemoor market – 'tas picked up a bit, but still tis a long way off what it should be'. The corn was coming along well and the lorries were busy, there was much to do. I had met few who worked so hard as the family here; even during the slower times of the year when the light is gone by five in the evening and the cattle are in the sheds, it was not unusual to see Tim in the workshop late into the evening, welding or mending, maintaining the lorries or some other piece of machinery by spotlight under cover of one of the old sheds. He was fortunate to live here on the Marsh, and many would no doubt be envious of the location and acreage that he owned, but he never took this for granted nor was he ever idle. In the forefront of his mind was always the well-being and continuity of the farm, the preservation of the land and the responsibility he had inherited from his father and grandfather and would in turn pass on to Jack. The continuing cycle of land management that had shaped these acres and many more across our countryside was played out here every day.

We weren't the only ones busy working in the heat of the evening. As Tim fired up the engine and the conveyor once again rumbled into life, I had been watching a spider

JUNE

who had ensnared a fly in her web which was stretched between two bolts that jutted out from the barn's uprights. The spider now busied herself, wrapping the insect quickly in her silken cocoon before cutting it loose and hauling it out of view behind a steel plate. We had almost lost the light by the time the deck of the trailer came into view, and had to complete the job by the beams of the tractor light. By the time the last of the bales ascended the conveyor, both Jack and I were sweat-soaked, stacking to either side of us and jamming the final bales into any remaining space. Eventually we crawled out from the barn, manoeuvring awkwardly between the bales in the half-light, and finding a foothold on the conveyor to climb back down to the yard. We had done a good job, each bay was now completely full and the hay rose tightly packed and straight-sided to the arched roof of the barn.

With haymaking complete I entered what, for me, is the slowest time of the year. There were jobs still to do on the farm that would keep me occupied and I was grateful for this, as days spent alone with little structure soon meld together and leave me feeling dissatisfied and depressed. Although I am happy living in a condition of relative solitude whilst concentrating largely on my work and day-to-day living, I have found that days spent without purpose, whether outside by the fire or within the wagon, do little to maintain a positive outlook. I am aware that for those who find little fulfilment in their daily employment the idea of being cut adrift from that relentless treadmill and spending hours relaxing by a campfire may hold great appeal. It is, however, something to be enjoyed sparingly; an evening spent under the stars whilst waiting for a kettle to boil for tea is undoubtedly one of the greatest pleasures, but it is enjoyed to its fullest when coming at the end of a good day's work.

If you observe the birds and creatures of the countryside, you will rarely if ever see them idle. They are always engaged in activity of some kind that is to either their or their offspring's benefit, ensuring their survival and well-being. It is rewarding to observe and to appreciate these seasonal habits of the countryside's other residents, but to appreciate it fully, one must be a part of it oneself. When engaged in one's own struggle with the adversities that life can throw at us, some comfort may be taken in remembering that the nature surrounding us also faces constant adversity, knowing that life is not easy for any creature. I find that this helps me find strength and purpose, leading eventually to a condition of stability and calm.

High summer was approaching, the long days were hot and often without breeze. At night-time I slept often with the door open as well as the window above my bed, attempting to draw any cooling night breeze through the wagon. I had to make sure that all lights within the wagon were extinguished before throwing it open, to avoid the living space filling with moths. The lights of the wagon, the only illumination here on the Marsh, would otherwise see them descend in great numbers. I was up shortly after five most mornings and would make tea and breakfast outside, greeting the rising sun as it emerged from behind the hawthorns to the east. By midday its scorching rays were directly above me, but the alder trees provided shade and shelter and the temperature under their leaves, where both the wagon and fireplace were tucked away, remained comfortable. The livestock on the farm also made use of the hedgerow and the trees that rose from it. During October and November, they had sought shelter there from the south-westerlies that whipped across the farmland; even the resilient cattle huddled together in its lee, grateful for the protection afforded by the wall of vegetation as it deflected the winds. Now during the high summer, the sheep lay in the shade of the hedgerow trees, sleeping through the intolerable midday heat, moving only when the sun's arc crept around to expose them once more to its searing rays.

I was keeping myself busy working at a farm that lay just beyond the village. Blackmore Farm was just a five-minute drive from my pitch on the Marsh and its farmhouse, an imposing Tudor manor, dominated the farmland beyond the village's cottages and church. The Dyer brothers had farmed this land alongside their father who had acquired the buildings and ground here in the fifties and then worked hard to build a dairy herd of enviable reputation in the following decades. Both the boys, Ian and Alan, had eventually married and raised their own families on the farm. The boys' father had reluctantly retired from his duties as he found himself unable to maintain his beloved cattle, the herd now much expanded from its humble beginnings, but the boys had continued their father's legacy, working hard to keep the farm both thriving and moving forward, although in order to do so, it had changed much since those early days. The farm still had a good head of cattle but these days it was beef animals that grazed its pastures and its milking parlour now stood empty; like many others in the district, the dairy at Blackmore had ceased milk production in the last decade. Although dairy still plays a role in the farming landscape of the West Country, it is now just a handful of big enterprises that make up that sector. These outfits are a far cry from the small dairy farms that were once dotted among the shadows of the Quantock Hills. Huge cattle sheds rise like agricultural cathedrals from vast acreages of grass and maize which will produce the silage needed to sustain the immense herds.

Those smaller farms that have managed to survive are now diversifying as best as they can in order to keep their gates open and maintain the land passed on to them from previous generations. Blackmore Farm was no exception. I

JUNE

had first met the Dyer brothers when coppicing an elm hedgerow on their land several years ago, when the milking parlour was still in operation. I had been greeted by Ian, the older of the two, dressed in a waterproof overall and apron of the type now little seen on the modern dairies. I had worked alongside the two brothers, cutting dead elms into cordwood which we then hauled back to the farm in an old dump trailer, to fill the cavernous fireplaces that heated the manor house over winter. I liked both brothers, but it was Ian with whom I would go on to have many dealings in subsequent years. As the farm diversified, Alan, the younger of the two, took the helm with the farming side of the business while Ian set about converting the manor to guest rooms which he let to the tourists who frequented the district throughout the year, as well as holding events on the property, as the manor provided a desirable historical backdrop to weddings and garden parties. Alongside this Ian also ran a farm shop which sold local crafts and produce. Despite these necessary diversions he remained a dairy man at heart, however, and the farm was in his blood. I had a lot of affection for him, a countryman and farmer who had both the keen wits and courage to forge his own path.

Between the manor and the farm shop, which sold both logs and kindling by the bag as well as by the load, I was guaranteed several weeks' work at Blackmore each summer, cutting and processing firewood. The wood was all suitable for burning, but was generally heavy and knotted and had been sourced from forestry yards who had taken the cream for their own wood fuel business before selling the more difficult lumps at a reduced rate. This arrangement suited both Ian and me; he had a saleable product that wasn't devoid of some profit and I was kept employed during my quietest time of the year. I cut in the cooler morning before retiring to the wagon in the afternoon and then returning when the heat had left the day. Ian generally dropped in several times throughout the day to check on progress and brought with him a cold pressed apple juice from the shop's refrigerator and often a cake of some description. His wife, Anne, baked many of the cakes and biscuits available in the shop and these, alongside her winter soups, are some of the best food I've tasted.

On returning from Blackmore in the evening I stopped in at the yard at Marsh Farm to fill my jerrycans from the tap just outside the workshop. The yard was quiet apart from the chatter of sparrows who flitted around the buildings and newly filled hay barn, and an occasional lowing from the calf shed though most of the livestock, both sheep and cattle, were now grazing the pastures beyond the yard. As I got back to the wagon I heard the cuckoo calling from the hawthorns; he would soon be away from this place, travelling south once more to Africa.

JUNE

I lit the fire, stripping to my shorts and enjoying the cooling air whilst making tea. From the hills I could hear the sound of gunfire — clay shooting kept the sportsman here occupied while they waited for the autumn game season — and then the bells pealed out from the church in the village. Once the water had boiled, I filled the basin and washed outside whilst keeping an eye open for any dog walkers who might be out enjoying this summer evening, as I was. I had not yet been caught unawares whilst washing outside and rarely saw anyone here after dinner time. The Marsh at this time was mine, a quarter-mile or more from the farmyard and surrounded wholly by farmland reaching down to the River Parrett that marked its boundary. After washing I climbed into the wagon, snuffing the wicks on the candles before climbing into bed, leaving both door and window open to the night air. The pipistrelles were once more hunting silently around the wagon as I fell asleep.

12

July

It was quieter on the Marsh in the mornings now. The dawn chorus of spring had been subdued by the sultry atmosphere of high summer which made even the birds lethargic. The water in the ditches and drains was low and covered completely with a blanket of duckweed. There was barely enough to cover the heron's ankles and lately he often preferred to hunt in the brook where schools of roach gathered in the deeper pools and hung lazily in the tepid water. The heron stood at the edge of these pools, motionless and statue-like. At this time of year he preferred to hunt in the early morning, when the sun was yet to illume the brook's depleted waters, skulking close to bankside vegetation where he might go unnoticed. His searching, reptilian eye took in every movement within the pool; the gentle fanning of the roach's fins, the rise and fall of each fish as they gulped air from the surface, each twitch of the shoal, he noticed. Waiting for the current, although weak, to edge them ever closer to where he lurked. Before the sun had climbed above the willows he had eaten and flown on, beyond the Parrett, to ditches and drains unbeknown to me that enclosed other cider orchards and pastures across this ancient landscape.

The long days were slow and I spent them, it seemed, at an increasingly leisurely pace. The firewood at Blackmore had been processed and bagged, and here at Marsh Farm I had completed all of the tasks asked of me. Hedgelaying would begin once more in six weeks' time, when the mornings dawned with a seasonal chill and the mists once again cloaked the ditches and wetlands. The great skeins of geese would begin to fly in from the channel to graze on the pastures alongside sheep and cattle. For now, however, I filled my days largely in recreation though I was rarely indolent, busying myself about the pitch engaged in some craft, reading in the shade of the alder trees, or observing the habits of the birds and animals who lived on the Marsh alongside me.

The hedgerow next to the wagon was quieter now, although the low drone of the bumblebees could still be heard. This weightiest of winged invertebrates seemed to defy gravity in its somewhat ungainly work routine as it careered between the flowering blackberry, thistles and

JULY

ox-eye daisies that grew close to the orchard. An iridescent damselfly strayed from the ditches and came to rest on the hawthorns by the wagon. The thorn trees were by now decorated with green and unripe berries that awaited the turn of the season, when they would turn a conspicuous dark rouge.

Along the hedgerows, the poppies were growing in great profusion and their delicate, papery petals turned patches of the field a brazen scarlet. The bees worked amongst the flowers and the thistles, whose royal purple flower head stood protected in the midst of the cluster of spiny leaves. The thistles were abundant around the old hedgerow that enclosed the orchard and I cut several of them with my pocket knife and arranged them in an old jam jar that sat atop the stove. It felt a long time since I had lit a fire in the wagon, and the thistle flowers lifted my mood just as the fire did during the colder months. The wagon's interior was filmed with dust blown in on the dry July winds, but with the doors and windows closed the temperature inside became intolerable, and the dust was the lesser of two evils.

July is the year's end for me, a time to rest before preparation for the hedging season. During August I planned to travel once more to Dorset to collect the hazel stakes I had cut back in May; they were still stacked in the coppice and awaiting collection. I would also travel to Earnest in Hampshire to collect binders for the second half of the Eweleaze contract, eight hundred in all. By that time the corn harvest would be under way across the southern counties. Here in Somerset, on many of the arable farms, the barley had already ripened and the whiskered seed heads were now heavy with grain and drooped under the July sun. Soon the combine harvesters would be cutting the barley before moving on to the wheat next month. The corn here beyond the wagon was starting to ripen slowly, though remained weeks behind the barley. The green wheat ears were beginning to take on the more familiar yellow-gold in the patches that saw the most sunshine, away from the shade of the alders and other trees that grew from the hedgerow that enclosed the crop.

On the cusp of harvest time, the countryside waited to see what the coming weeks would bring. Despite the application of synthetic chemicals to ensure the crops had grown unhindered by weeds or pests, on artificially enriched soils that saw the crop grow vigorously, regardless of whether the earth had been rested or not, it would still be nature who had the final word. There is as yet no way to control the weather, which still holds the power to thwart the year's work if adverse conditions render harvest impossible. The farmer's work hangs in the balance now, an annual poker game in which he is all in, wagering everything on this small window of opportunity in which success is far from

JULY

guaranteed. The farmer has always been a gambler, pinning his hopes on fair weather and a good crop, and although modern technology sees him dealt many aces, nature remains the dealer.

It was the radio that informed me of a change in the weather to come; the oppressive heat that had hung across the Marsh for weeks was set to break within days. I was relieved to hear the forecast; unlike the cold of winter, the summer heat is inescapable in the wagon and saps both my energy and my enthusiasm. Despite the many benefits of the fair weather, I found myself starting to crave the crisp mornings of autumn, that comforting and familiar chill, and the company of the hedgerow. The farmers here had been keeping a close eye on the weather and now it was time for them to play their hand. First up was the barley, which would not keep as well as the wheat. Each seed head nodded lazily under the July sun, and each stem now grew increasingly weak. If allowed to dry for too long the barley stem will fail, becoming brittle and increasingly likely to snap. The heavy summer rain this close to harvest could see many of the seed heads lost or patches of the crop flattened altogether. When this happens even small pockets of lying plants will be exploited by the wood pigeon who will descend to feed on the spilled grain and likely flatten yet more of the crop in the process. Unlike the barley, the wheat does not drop its head and instead remains upright, so is less likely to be damaged by any adverse summer weather, affording the farmer a little more leeway when considering a later harvest. I was unsurprised, therefore, to see the familiar dust clouds reappear, cast upon the air by the combine harvesters which now began their work beyond the Marsh. The opportunistic wood pigeons immediately descended in flocks in the machines' wake, frantically filling their crops with the grain spilled by the farmers' activities.

As the light faded on the Marsh I sat by the light of the fire, accompanied by the now-familiar pipistrelles who always joined me on the warm moonlit evenings. The harvest however continued unabated, and even in the silvery half-light the dust continued to rise conspicuously from the barley fields, the vast plumes illuminated by light beams that moved in all directions. The tractors, emblazoned with white light, shadowed the combine so that once it had taken its fill the enormous machine could discharge its haul into their waiting trailers. Neither combine nor tractor would stop during this manoeuvre and the operation is carried out on the hoof, the cutter bar of the combine scissoring the barley stems mercilessly and drawing the grain into its capacious tank even as it empties its current load.

It was gone midnight by the time I climbed the oak steps into the sleeping compartment of the old wagon; the air hung thick and close under the alders which made it

JULY

uncomfortable, and despite the late hour I found it difficult to sleep. As the church bell struck in the village, I was still alert enough to count the chimes ringing out twice, muffled and low but reaching me in the wagon nonetheless. I lay awake, considering that in little more than three hours the sun would already be lightening the sky to the east. Eventually, and to my relief, the hour began to get the better of me and despite the stifling heat, I settled. All the while, the distant sounds of the combine harvesters' rhythmic cutting showed no sign of abating. As I drifted off, I spared a thought for the men harvesting the barley, who would have little sleep until the storms arrived in the coming days.

The barley straw had been baled and hauled off the fields by the time the weather broke. The hayfield on the Marsh had grown little since being mown last month and was looking parched, the cut brown stalks not dissimilar in colour to the freshly cut barley fields. With the barley safely in, the storms would now be welcome; if the fields could be reinvigorated with summer rain, more grass could be cut for silage on the dairy farms and the maize crop, due for harvest in the autumn, would benefit greatly. The pastures, too, needed the rain to sustain the livestock through the last of the summer. All around, the country was thirsty and although the familiar template of enclosed fields was still outlined in the bold green of the hawthorn and elm hedgerows which divided them, the patchwork was predominantly beige.

The weather approached from beyond the Quantock Hills. The sunlit barley stubble and wheat fields on the hills' northern slopes contrasted sharply with the heavy, brooding sky which approached from the south-west, and the swallows flew excitedly about the orchard as if heralding the event about to unfold. I stood outside in the thick, oppressive air and awaited the rain. I sipped my tea and watched the bumblebee still working, seemingly unconcerned by the approaching storm.

I managed to stow the firewood beneath the wagon before the rain came. The ever-darkening sky now hid the sun completely from view and the countryside all around fell eerily quiet. The swallows were now noticeably absent from the air and the magpie had retreated to the orchard with only the occasional nervous chatter; all knew instinctively that it was time to seek cover. The thunder rumbled menacingly from beyond the hill although just the slightest breeze rustled the leathery green leaves of the alders. As the rain began to fall, I, too, retreated, drawing back into the wagon's interior, though leaving the door open and secured on its latch. The atmosphere was still muggy and I was dressed in only a pair of shorts as I settled into the comfort of the armchair to enjoy the increasingly unsettled weather.

It was not only the birds and myself that hunkered down against the weather; the livestock on the farm would also

JULY

be seeking shelter beneath the hedgerows where they would be shielded from the worst of the storm. Hedgerows provide more than just a habitat for our farmland wildlife, more than arboreal highways connecting woodlands and certainly more than rural boundary markers preserved for aesthetic value alone – they also have great value to the farmer. Lamb survival rates in the springtime are increased when the shelter of a hedgerow is available, reducing the effects of windchill and hypothermia during the inclement early spring weather. Then, on the hottest summer days, hedgerows and their trees provide shade and prevent animals becoming overheated. Heat stress in livestock leads to reductions in milk yields, growth rates and disease resistance.

As I watched the rain falling, I was also reminded of the role our hedgerows play in flood prevention. Flooding has become ever more common in recent years amongst a marked increase in dramatic weather events. The ground in this district had been baked hard by the hot weather over the recent weeks and the water would now flow across it as if it were concrete. I knew the lanes on the hills would be awash as the water ran freely from the freshly cropped barley fields and from the wheat and maize fields still to be harvested. The run-off, stained a distinctive red ochre by the Quantock soil, would flow like a river and quickly overwhelm the drains before running unhindered into the villages and hamlets nestled amongst the farmland. Before this floodwater reached the sandbags being hurriedly propped in the cottage and pub doorways by frantic residents, it would be the hedgerows that formed the first line of defence. The hedges and trees slow the floodwaters and allow more time for it to infiltrate the ground, as well as giving the residents in the villages more time to prepare. The roots of the hedgerows run deep, much deeper than the roots of barley or corn, and will help the soil to absorb more water. Once the water is trapped the hedgerow will store it, holding it like a sponge; it is this ability that sees the hedges remain lush and green when much of the surrounding country seems parched and dry.

Hedgerows also help retain the valuable topsoil that is otherwise swept up in the floodwater. Eventually this earth-stained run-off reaches our watercourses; our ditches, brooks and rivers. It is here that the precious and fertile topsoil is deposited, lying as silt in the river beds, sometimes in such quantity that it hinders the spawning of our game fish, which rely on the silt-free gravel of our becks in which to reproduce. Of course, any chemicals that have been applied to the soil will also end up in our watercourses, and the fertilisers that artificially enrich the earth cause aquatic weed growth to flourish, unnaturally accelerating the plant growth in a similar fashion to our farmland crops and causing it to overwhelm the ditches and drains. This then further exacerbates the flooding problem. Our activities on the land have far-reaching consequences.

JULY

After the storm had passed the air felt refreshed, clearer and considerably more comfortable. There had, however, been a casualty here on the Marsh: one of the old willow pollards had split and part of its crown now lay strewn across the track, its hollow trunk torn in half and its clean white innards exposed to the sun's light. Tim and I eyed the casualty and considered the clean-up operation before he generously offered me the wood for my own use or to sell. I had already processed and stacked Tim's winter fuel wood, and his log store contained much harder and hotter-burning woods gathered from previous windfalls about the farm, such as oak and ash, so on this occasion he was not interested in the willow logs, although they would burn well enough once suitably seasoned.

Tim moved the heavy bough with the tractor. As it dragged the crown away from the track, clearing the access to the Marsh, the last sinews that connected the fallen crown to the main stem were severed. These living links are how the willow tree multiplies; its habit of collapsing and its willingness to then take root from any number of fallen boughs or branches has seen it colonise the banks of many of our rivers and watercourses. Several species of willow are cultivated on our wetlands for the weaving rods it produces and of course its straight-grained, soft and easily workable timber which amongst other things is still crafted into the bats for that most English of rural pastimes: cricket.

The willow is one of many trees that can be pollarded,

cutting the tree's crown during the winter months so that it throws up fresh rods that can be harvested. In the same way that coppicing prolongs the life of many woodland trees so, too, will pollarding increase longevity, the ancient rootstock giving rise to vigorous, juvenile growth which provides a sustainable source of wood fuel or building timber. For centuries hedgerow trees have been managed by pollarding, their position on the sidelines amongst the hedgerow meaning that they could provide a sustainable energy supply for the custodians of the land, while the grassland remained free for pasture or cultivation. With trees at different stages in their growing cycle an annual crop of wood fuel could be guaranteed, and maintaining the crowns in this way also meant that they never shaded out the thorns below, ensuring that the hedgerow's primary function as a stock-proof fence was never compromised. The correctly managed hedgerow is still capable of providing this fuel, with the trees growing within its length serving as habitat until the time comes to harvest the wood of the crown and renew the life cycle of the tree once more.

I spent the afternoon under the willow pollards, first cutting the branches into logs and then stacking them neatly at the base of the old willow's fractured stem. I then retrieved several old pallets from the farmyard which I set alight next to the trees to create a hot bed of embers on which I could torch the brushwood, which was full of moisture and would otherwise struggle to burn. A gentle breeze fanned the fire

JULY

and encouraged the flames; the temperature had dropped by over ten degrees since the storm had passed through and although still pleasantly warm, the air was now fresher and caused the willow smoke to roll away quickly between the stems of the pollards and out across the hayfield, which had itself already taken on a fresher shade of green. The willow brush burned rapidly, and I raked the embers into a neat pile with an old pitchfork before nestling the blackened kettle amongst them.

I leaned on the fork whilst waiting for the kettle to boil and let my eye follow the smoke away across the meadows. This landscape, like most on our islands, has changed over the last century, with miles of hedgerows lost along with many of the cider orchards for which the county is famed, but it retains a great beauty and echoes of those former years. While the remaining hedgerows may have declined, they have yet a role to play in matters that reach beyond the farm gate, indeed in matters that stretch beyond its immediate locality and ultimately benefit not just the local environment but the United Kingdom as a whole, and perhaps even further. I have already touched upon many of these, but there is one last role that has the most far-reaching consequences of all. Hedgerows, and especially those which contain trees of some maturity along their length, are now being recognised as an invaluable tool in helping us reduce the amount of carbon dioxide in our atmosphere. In a nation that has comparatively little tree cover compared to our

European neighbours, these woodland arteries store carbon both above and beneath the soil. Carbon is stored under the leaf litter and in the hedgerow's roots while above ground, the hedgerow and its companion trees act as a sponge both absorbing and storing carbon. The remaining network of hedgerows, then, finds new relevance and purpose for not only our rural but our national community, and is an asset which we can ill afford to be without.

Lying on an old blanket beside the fire under the willows, I looked skyward. I counted four buzzards riding the thermals, their outstretched wings static as they captured the warm air which rose now from the Somerset flatlands. Though they were soaring at considerable altitude, I could see still the birds' broad, splayed tails moving left and right, feathered rudders that held them on an even keel as they circled, piercing the quiet with their shrill calls. A pair of rooks caught sight of the raptors and croaked disapprovingly. These two wind-beaten scouts looked like veterans of the rookery and were both dull and unkempt, their weathered and untidy appearance matching their guttural croaking. They often patrolled here and harassed the buzzards whenever the chance presented itself. On this warm July afternoon, however, neither seemed able to muster the enthusiasm to give chase and the buzzards circled ever upwards and out

JULY

of reach, calling victoriously to one another as the rooks disappeared from view.

Beyond the orchard, the cornfield over the hedgerow had now completely turned, and soon the combines would be rolling again. The summer wind pushed through the wheat and the crop waved gently, the swollen seed heads rustling together and startling a small flock of pigeons who took off with a clapping of wings, swerving sharply as they caught sight of me walking in the orchard. The apples were forming on the orchard's ancient trees now, and the wasps already flew between them, seeking the damaged or rotten fruit which might give up its sugars. The wasps had nested close to the orchard gate and remained calm unless provoked, but recently they had taken to visiting the wagon where they gnawed at the wooden window frames. If resting in the sleeping compartment of the lorry during the day I would often hear them, their sharp jaws scraping the hardwood, loud enough to prevent me from napping.

The hayfield continued to green and the cooler weather coupled with the occasional summer rainfall had seen it rejuvenated and the grasses now grew again unabated. Tim would bring the sheep here soon. The cattle tended to stay in the fields closer to the farmhouse while the sheep would graze the farm's peripheries. The sheep would cause me little concern – generally they are less inquisitive than cattle – and there was plenty of shade along the hedgerow's length to provide them with shelter. In the past, when stopping in more exposed locations, I have been accompanied by sheep jostling for position under the wagon, the low bleats of the ewes accompanying the crackle of the fire inside and the rain falling on the wagon's tin roof. I don't mind, the only annoyance is the flies that inevitably shadow livestock and find their way in through the wagon's open doors. But this is a minor inconvenience, only to be expected when living on the farm, and the livestock also bring with them the swallows, flying joyously above sheep and cattle to snap up the insects.

As the sun began to descend in the west the high cloud that had gathered over the course of the day started to reflect the crimson light. At first just a blush of colour, by the time I had made my way from the willow pollards to the orchard the sky was ablaze. An unbroken flaming horizon stretched out before me, spanning from the southern hills to the flatlands away to the north, fading to the east as the evening firmament turned an ever-darkening blue. The first bright stars were already visible in the darker eastern sky and the pipistrelles were on the wing. On returning to my pitch, enough light remained for me to break off some of the dead elder sticks in the hedgerow and light another fire outside the wagon – the lengths of firewood stowed underneath it were dry and seasoned and saw the flames build

JULY

rapidly. I retrieved a bottle of cider from inside before taking a seat close to the flames.

My thoughts wander in the firelight. Increasingly these days I find myself contemplating what my role is, and what my obligation to the hedger's craft involves. It is now approaching thirty years since I sat beside another fire, in the company of Bill Bugler, and listened intently to both his musings and his instruction. I was a naive young man back then, enchanted by those first weeks in the coppice, but although much water has since passed under the bridge, I remain just as enchanted by our woodlands and hedgerows today. My life, like everyone's, has included its share of stress and anxiety, and I have been greatly comforted during those times of upheaval by the countryside that is available to all of us. I have found that my bond with it has been perhaps the most rewarding of all my relationships, and I have taken great pleasure in being able to introduce its generosity to my daughters and show them its beauty. To my delight, they have been as enchanted as I am. As I grow older, I find I rely on this relationship ever more, not just for my employment but also for my sense of place and purpose. My role, I have concluded, is to pass on the skills I have gathered over these last decades: the hedger's craft.

Hedgelaying was all but forgotten in the latter half of the last century, an outdated technique for managing outdated boundaries in a landscape that apparently needed new ideas to see its potential maximised. Old knowledge was largely discarded and practitioners of traditional methods of land management were scorned as a new agenda was put in place. The final decades of the twentieth century saw a ruthless march forward that seemingly cared neither for craft nor tradition, for wildlife or heritage, and sought only an increased output that eventually saw the countryside run along similar lines to a city business, overtaken by short-sighted greed: factory farming. Some country people flourished, but many didn't and were instead cast to the wind, their legacy overridden by developers who purchased newly vacant farms and cared little for their past. The countryside has been shaken, transformed over just two or three generations into a quicker-paced, more industrial environment that, although still retaining much of its charm, has begun to diminish both culturally and ecologically.

But they still persist, those men and women who eke a humble living from its fields and woodlands, who remain as tied to this environment as the very trees and hedgerows themselves. And it has been these countrymen, these indigenous residents of the West Country who live and work by the seasons and who have found abundant wealth in a life devoid of much financial reward, who have come to shape my own outlook on life. The wisdom they possess is not taught in any school, but rather passed on in woodland and farmyard, in village inn and hay barn. The demise of traditional mixed farms and heritage craft, alongside those people who have preserved and passed on the skills that saw our

JULY

rural environment managed sustainably and sympathetically, is to be counted alongside the countryside's diminishing biodiversity.

I have worked amongst these people, and within the hedgerows and woodlands of the south-western counties of England, for more than a quarter of a century, and during this time have listened willingly to advice and guidance delivered to me most often in soft West Country tones by men whose years and experience have exceeded mine by several decades, custodians of our countryside who still practise skills that have been forgotten by many, but who now find themselves conservationists by default. Their work continues to preserve our rural heritage whilst also benefiting species diversification in the countryside, and so is beginning to find new relevance in a changing world. Change is inevitable, as well as necessary, but amongst these men who still live and work as their fathers before them did, who have stubbornly refused to be swept along by a tide of modernisation that saw much of what they knew to be good disregarded, I have found much wisdom and integrity. Now, perhaps, it is my turn to pass on what has been given to me.

The tide is, in some ways, starting to turn once more, and the surviving hedgerows are thankfully again being recognised for the services they provide our countryside, finding new relevance in an ever-evolving landscape. The hedger too, must continue and evolve, and my hope is that, in an age that sees success measured largely by the accumulation of material wealth or social status, there are still many amongst us that find pleasure and reward not in such things but perhaps, in a greater sense of purpose. Although likely going unnoticed by many, the humble work of the hedger will continue to have far-reaching benefits for both farmland wildlife and rural communities and indeed for the very land itself. It is then these people to whom I seek to pass on the hedger's craft and ensure, as best as I am able, the hedgerow's survival.

It is a most beautiful location here, and indeed I have stayed in many such idyllic spots over the years that I have spent travelling these districts and working the hedgerows. But I would not be here much longer. Like the swallows gathering on the telegraph wires in late summer, anxious to start their migration, I too was keen now to be away. Although this place had been home now for months and in many ways I had come to feel a part of it, once the corn was off the fields, I would be gone, leaving only the ash from the fire to tell of my stay.

The autumn and winter months would again see me travel between all of the counties of the West Country and already I had secured pitches for the old wagon on them all. Although each of these south-western counties has its own unique attributes and personality, I have found the people from each are similarly warm and hospitable, and all

JULY

those that remain truly tied to the land have a great fondness still for the heritage and traditions of this peninsula. I have greatly enjoyed the wild and untamed north Cornish coast, catching sea bream and bass from its Atlantic shores and teaching my children to swim in its rivers' calmer estuaries. I have walked Exmoor's ancient hedgerows and burned charcoal in Devon's lesser known woodlands. With each year I have welcomed the spring, whose first subtle victories over winter's hold are won here in England's quiet western backwaters. I have walked the high chalk of Wiltshire, laying hawthorn and field maple across its ancient downland and returning in summer to watch the blue butterflies dance amongst its flowering meadows. It is in Dorset, however, that I have spent the majority of my adulthood, passing many a summer hour on the beaches and rocky outcrops of the Jurassic coast. Beneath weathered cliffs and on sun-warmed pebbles, my daughters and I have fished and slept, watching the fireflies in July decorate the chalky hillsides that rise inland from Dorset's clear coastal waters. It is in this county that both of my daughters were born and where first this long journey through the woods began, where first I was inspired, and where my love of the hedger's craft was first conceived.

Even during those settled years in Dorset, I spent many months away, engaged always in woodcutting and farm work. I often felt restless, and was at my most contented when working amongst the trees or when seated by the fire, and even now with several decades behind me, it remains this way. When I left the cottage in Dorset some years ago, it was not without hurt, but the road leads ever onwards and opportunity lies along its length. Rare is the day that nature has not offered me a reward for continuing forward when at times I have felt weary, and it is this steadfast relationship that sustains me when others have faltered. I am thankful, having encountered new love and friendships while retaining older bonds, and feel that the road ahead holds still more, which in time will reveal itself. But although my journey will continue, I feel the time is approaching to settle once more, to begin a new chapter that I hope will see others come to work amongst the hedgerows and woodlands.

I have long desired somewhere of my own, a piece of ground here in the south-west where I can put down roots and lead a more settled existence, and I have been working towards this end for some years. My aim is to continue to live simply on the land and be able to use this place as a base from which to pass on the rural skills I have learned. I have moved often throughout my life; I left my place of birth in eastern England at the age of four and lived abroad for the greater part of my childhood, growing up on the North Island of New Zealand, where I spent much time in the wild places the country is renowned for. It was here,

JULY

amongst the country's farms and woodlands, that my love of the outdoors was first ignited. On returning to England at the age of fourteen I spent just four years on its eastern coast before, at eighteen, when I became able to look after my own affairs, I left and took the road west. Here I sought employment on West Country farms until the road led eventually to Dorset and my journey as both a father and a woodsman began.

Time has passed quickly, and as my fiftieth year looms ever closer, I think more often of a place of my own and what I could do with it. As I have aged, my desire to teach has grown, and although I never imagined myself in that role, it now holds great appeal for me. I left school with no qualifications and went straight into work; the men and women who attempted to teach me had little time for a boy who desired so strongly to be elsewhere. It wasn't until five years later, amongst the understorey of a Dorset woodland, that my education truly began. It was here I learned not just of wood and trees, not only of heritage and tradition, but of what I consider still to be perhaps the most valuable lesson of all. It was here that I learned a sense of place, that like the other inhabitants of the woodland I, too, was a part of the land's story and my life, like theirs, was governed by the seasons. I learned that my role as a custodian was to make sure I took no more than was necessary and that I should respect and repay the generosity nature had bestowed on me. I learned that people will come and go and that our relationships with each other are ultimately finite, but our collective relationship with the land stretches back beyond imagination, and like any other needs to be maintained and cherished.

Now has come the time that I wish to give something back, something more than just the labour I have been able to offer thus far. Although I hope I have still decades to work and live on the land, my ambition is to be able to welcome people to my own ground and to touch on the things that were so generously passed on to me as a young man in Dorset. I envisage a dwelling of some kind where I might live in a little comfort, and the wagon could then serve as accommodation for visitors who had perhaps come from further afield. Here I would teach the methods of hedging, of coppicing and weaving, while still working on the hedgerows during the winter season. Although it has taken much time and has often felt out of reach, with each passing season, I progress closer to this goal. My hope remains that some will be persuaded, as I was, that a life in the hedgerows, amongst the fields and the woodlands, the thorns and the briars, is one of the most rewarding of all.

Acknowledgements

Thank you to my agent, Sarah Ballard of C&W Literary Agency, whose encouragement and faith in my writing inspired me to expand on the rough notes and jottings I first read to her. Your guidance and experience have been invaluable.

Thank you also to my editors Kris Doyle and Kat Ailes, who have guided me patiently and considerately through a process of which I knew very little and who have honed my writing. The book is better for your work.

Many thanks to the friends and companions who have supported me along this journey and offered both support and advice: Tim and Susan Popham at Marsh Farm. Ian and Ann Dyer at Blackmore. Paul and Gemma Hackman, Ben Short, Liz Newall and Rachel Hall at Gutchpool Farm. Philip and Lisa Bell at Frogmore and James and Charlotte at the Story Pig.

I must also thank Joan Buck, who regularly put aside the

antiquarian nature books that arrived at her little bookshop in Wadebridge in Cornwall, and passed them on to me. I enjoy them still.

Finally, I must make special mention of my family: the support they have given me throughout this journey has been unwavering and I am eternally thankful for their kindness, support and encouragement.